THE GREENHOUSE BOOK

THE GREENHOUSE BOOK

SIDNEY CLAPHAM

DRAKE PUBLISHERS NEW YORK

Published in 1976 by
Drake Publishers Inc.
801 Second Avenue
New York, N.Y. 10017

ISBN: 0-8473-1347-6
LC: 74-22566

Printed in the United States of America

Contents

List of Illustrations

PLATES

IN TEXT

The photographs on pages 33, 34, 51, 52, 69, 87 (upper), 105 (lower), 106, 123, 159 and 160 are by J. E. Downward

The remaining photographs are from the author's collection

Line drawings are by W. G. Clapham

Chapter 1

The Greenhouse in General

In recent years the term 'greenhouse plants' has become something of a misnomer, as many of the exotic subjects covered by it are commonly grown indoors as well as in the greenhouse. The widespread use of central heating, large windows, conservatories, 'garden rooms' and other glazed extensions to the house has made it possible to grow many of the tender plants which need more heat than is usually found in the average greenhouse; in fact at one time these plants were considered suitable only for the 'stove house', a greenhouse kept at 70° F (21° C) or so.

Although this book deals mainly with the greenhouse, it is hoped that much of the information in it will prove useful to those who have no greenhouse as such but who nevertheless find great pleasure and enjoyment in growing 'greenhouse plants' indoors.

In its widest sense the term 'greenhouse plants' must cover also the less decorative but more useful crops such as tomatoes and other fruits and vegetables that can be grown in the greenhouse, together with those flower crops that spend at least part of their life under glass, such as chrysanthemums, carnations and bedding plants. Unlike the purely decorative subjects these are essentially plants for the greenhouse rather than for indoors, so they must have a place in this book, particularly as many greenhouses are largely devoted to them.

Staging

One thing that will have to be decided when the greenhouse is first obtained is whether or not to have benches or 'staging' in it, although this will be determined mainly by the type of plant that is to be grown. If for instance tall plants such as tomatoes and chrysanthemums are to occupy the greenhouse for most of the year, the staging can be dispensed with; or at any rate it will be necessary to have only a temporary arrangement that can be erected as required.

For most ornamental plants grown in pots, however, staging is almost essential, not only to bring the plants to an easily workable level but also to keep them near the light. The usual practice is to have staging down both sides and across one end of the greenhouse, but if tomatoes and chrysanthemums are to be grown as well as pot plants, it may be used on one side only, leaving the other for the taller subjects grown in the ground. Whichever arrangement is used, the staging should be not more than 3ft wide and if it is placed on both sides of the greenhouse ample room should be left for a central path.

Manufacturers of greenhouses usually also supply suitable staging for each individual model, generally in the form of a bench with stout slats set at intervals across the top. This is probably the best form for use in winter as it allows warm air to circulate up through the staging and round the pots. In summer, however, better results will be obtained if the top is covered with asbestos sheeting, so that a layer of sand, peat or gravel can be placed on it and kept moist, thus creating a more humid atmosphere around the plants.

If the greenhouse is a reasonably large one a very attractive effect can be obtained by arranging the staging in tiers, so that the plants stand at different levels, with possibly some trailing ones to hide the actual staging. This is a good arrangement too in conservatories and garden rooms, as apart from the fact that the plants look more decorative, many more of them can be grown without taking up too much space.

Heating

To make the most of this wide range of ornamental and useful plants some artificial heat is essential, for although most plants can be grown satisfactorily without it from late spring to autumn, the choice is so limited for the rest of the year that the greenhouse is practically useless then. On the other hand, if there is sufficient heat to maintain a temperature of 45–50° F (7–10° C) in winter, which is enough to convert a cold greenhouse into a cool one, it may be put to good use all the year round, for many of the commonly grown ornamental plants need no more than this. It also enables an earlier start to be made with tomatoes, bedding plants, chrysanthemums and so on.

With a higher winter temperature still, say 50–60° F (10–15° C), and with the help of a propagating frame that can be kept even warmer than this, the scope is again increased enormously. But beyond that the cost of heating the greenhouse to stove house temperatures becomes prohibitive for most people, and the comparatively few plants which need this sort of warmth can be grown just as satisfactorily indoors.

Of the different types of heating, electricity—generally in the form of tubular or fan heaters—has much to commend it: it is clean, labour-saving, readily controllable and easily adapted to take additional items such as soil-heating units, mist propagators and lighting. Its great drawback, apart from the possibility of blackouts due to power cuts, is that it is fairly expensive, particularly if fairly high temperatures are aimed at. As a rough guide the consumption is almost doubled with every 5° F (3° C) rise in the temperature to be maintained, so that for instance it would take nearly twice as much electricity to keep the greenhouse at 45–50° F (7–10° C) as it would to keep it at 40–45° F (4–7° C).

For small greenhouses that are to be kept merely frostproof, paraffin heaters are commonly used. Though not ideal, these will do the job satisfactorily as long as there are no noticeable fumes from them and adequate attention is paid to keeping them clean and filled.

For larger greenhouses the traditional system of a boiler and hot-water pipes is still as good as anything, and if the boiler is heated by gas or oil, the labour involved in attending to a solid-fuel boiler will be eliminated.

Aids to propagation

Whichever heating system is used, it is advisable to have a warm propagating frame as well to provide an extra high temperature for those seeds and cuttings which need it. Small electric propagators, suitable for use either indoors or in the greenhouse, are readily available; or on a larger scale a frame can be made over an enclosed section of the heating pipes or over a bed of sand heated by low-voltage electric wires from a transformer—a very economical method of soil heating. As the frame is intended to create a 'close' or warm moist atmosphere for the seeds or cuttings, it should be covered with a fairly close-fitting glazed light so that ventilation is completely under control.

A further refinement where the taking of cuttings is concerned is a mist unit, which automatically provides the necessary moisture and humidity. This is normally used in conjunction with a soil-heating unit which supplies 'bottom heat' when the cuttings are inserted in the sand-bed, while overhead moisture is provided from a water-supply which feeds a series of nozzles capable of ejecting a fine mist. These nozzles are controlled by a sensitive device generally known as an 'electric leaf'. This automatically switches the mist on whenever the atmosphere reaches a level of dryness corresponding to the condition of the cuttings, and off again as soon as these have been covered with an adequate film of moisture. Although used most extensively for commercial purposes, systems of this sort are available to the amateur gardener.

CHOOSING AND SITING THE GREENHOUSE

With heating being as expensive as it is, the main thing is to avoid wasting it—an important factor, among others, to be taken into consideration when one is buying a greenhouse. As glass is an excellent conductor of heat, the more there is of it the greater will be the heat loss, so that unless mainly tall plants such as tomatoes and chrysanthemums are to be grown in the ground a greenhouse with side walls of brick or wood is preferable to one with glass almost to the ground.

The lean-to type of greenhouse, preferably backing on to a south wall, will also save a certain amount of heat: the rear wall not only protects the structure from the north wind but also acts as a mild storage heater after a sunny day. But this apart, the lean-to is far from ideal, because the plants will grow away from the back wall towards the light and can soon develop a one-sided appearance.

The span-roof type of greenhouse, with its roof in the form of an inverted V, is in any case far more commonly used than is the lean-to. To avoid wastage of heat, it should if possible be placed where it will be protected from the prevailing wind. Otherwise, there will be constant draughts that will soon bring the temperature down, particularly if the ventilators do not fit too well.

As for the greenhouse itself there is an enormous selection to choose from, the two main types being the traditional wooden span-roof one and its equivalent in aluminium alloy. The former, if well constructed in the first place and kept well painted afterwards, will last for many years and it perhaps looks the more attractive of the two; but the aluminium alloy type has the advantages of needing less maintenance and admitting more light—an important factor, particularly in winter, when the plants need all the light they can get.

With both these and other types, consideration should be given also to the number of ventilators, for most small greenhouses have too few to cope with the sudden rises in temperature that can occur in summer. The preference should therefore be for those types best equipped in this respect.

The actual siting of the greenhouse also needs more consideration than it usually gets. The site, as well as being protected from the wind, must also be fully open to the sun, so the greenhouse should not be placed near trees or large shrubs likely to obstruct this. The common practice of putting it close against the boundary in the bottom corner of the garden is not a good one either: it is often difficult to get round it for maintenance and there is always a tendency to neglect the one side and end that are out of sight. Such a position is likely to create difficulties too with supplies of electricity and water, whereas there will be little trouble with these if the greenhouse is sited near the dwelling; in any case it will be more convenient to have it there, particularly in winter.

Chapter 2

Various Requirements

Apart from the major items already mentioned there are a number of other things that will be needed, but most of these are best kept in a garage or garden shed. Otherwise, by cluttering up the greenhouse, they not only make work more difficult but also provide shelter for slugs and other pests. It is, unfortunately, only too easy to accumulate dirty pots and seed boxes, bags and bottles of fertiliser and other greenhouse paraphernalia on and beneath the staging.

The potting bench

Nowadays few gardens run to a potting shed as such and one is certainly not necessary for the small-scale work done by most amateurs. A small portable potting bench is, however, a very useful item and is easy enough to make.

Basically the bench should be at least 3ft long and 18in wide, in the form of a tray with the back and sides about 6in high and a front board no more than 2in high: if it is higher than this it will make working difficult. No legs will be needed, as when it is in use the bench will be merely placed on the staging, but a covering of linoleum or plastic material over the base, which should be made strong and rigid by means of cross-battens on the underside, will ensure a smooth working surface.

Seed pans and trays

A supply of these will be needed both for the sowing of seeds and for the pricking out or transplanting of the seedlings (see Chapter 5), although for very small sowings ordinary plant pots can be used instead.

For the small quantities of seed that the amateur usually sows, it is best to use the small earthenware or plastic seed pans that are available in the same diameters as plant pots. One type of seed can then be sown in each, whereas if ordinary standard-sized

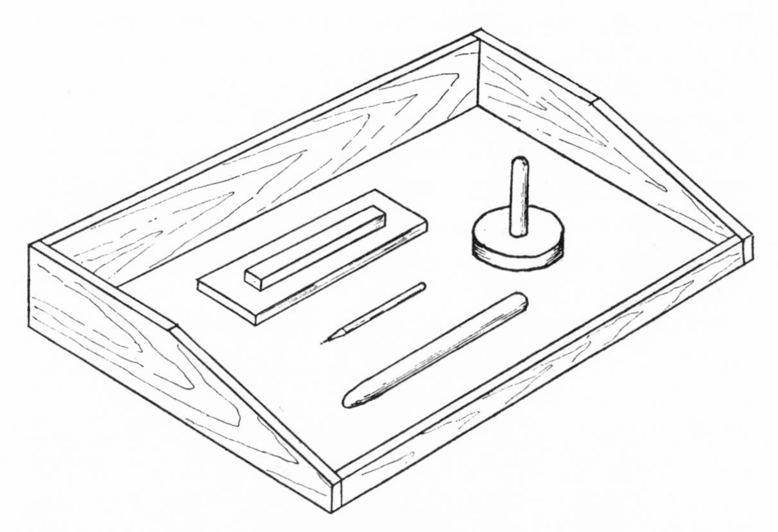

Fig 1 A simple portable potting bench, with a wooden presser for seed trays, a circular for pots, a small dibber for pricking out and a potting stick

seed trays are used—these are based on the commercial wooden seed box 14½in × 8½in × 2in or 2½in—there will be room for two or three different sorts, which can lead to difficulties when one sort germinates before another. A supply of the standard-sized ones is, however, necessary for the pricking out of seedlings, or if ordinary wooden boxes are used for this purpose, they must have in the bottom of them adequate slits or holes for drainage.

Whether the seed trays are of earthenware, plastic or wood makes little difference to the growth of most plants but plastic ones are lighter, cleaner and usually designed to nest into each other, which can mean a considerable saving of storage space.

Pots

Nowadays the traditional clay pot has largely been ousted by containers in other materials, notably some form of plastic. A good-quality plastic is the best for most purposes, although for short-term plants such as those started off in spring for summer bedding, the more temporary pots in paper, peat and other materials are quite suitable.

Despite the changes in the material used, plant pots are still made largely in the traditional sizes, the main ones being the 3in, 3½in and 5in ones (these measurements referring to the top diameter of the pot). For most ordinary, decorative pot plants these are almost the only sizes needed but for tomatoes, chrysanthemums, carnations and some ornamental plants a supply of larger ones, from 6in to 10in may be required.

As with seed trays, drainage of the pots is important. This is usually provided by crocks placed concave side downwards over the drainage holes. Where only plastic pots are used these crocks can be something of a problem, as when a pot is broken—and some types are unbreakable—the pieces are not suitable for the job. In this case proprietary crocks made of perforated zinc can be used instead.

Composts

To get the best results with seeds and plants it is advisable to use standardised soil mixtures or composts of which the basic qualities of good aeration, moisture retention, texture and food-supply can be reproduced almost exactly time and time again. The modern practice, therefore, is to use certain composts that are made to definite formulas and among these the John Innes seed and potting composts are perhaps the most popular. Originally produced by W. J. C. Lawrence and J. Newell of the John Innes Horticultural Institution, these are based on the use of partially sterilised loam and they come as near as possible to meeting most of the needs of seeds and plants. In addition, they are completely free from weed seeds, pests and fungus diseases.

They are available in four types, one for seed-sowing—the

J.I. Seed Compost—and three for potting, these being basically the same except for the amount of fertiliser in them. The idea of having these different amounts of fertiliser is not so much to improve growth in the early stages of the plant's life as to provide 'staying-power' in the compost. Thus, while the J.I.P1 will do for weak-growing or short-term plants, No 2 is needed for stronger growers and No 3 for very vigorous ones such as tomatoes and chrysanthemums.

All these composts can be easily obtained. Although they are not always of the high standard set by the originators, they will usually give satisfactory results. It is possible to make your own composts, but as most readers are hardly likely to go to all the trouble of obtaining and preparing the correct ingredients, I shall describe only the simplified procedure. Further information can be found in *Seed and Potting Composts* by Lawrence and Newell, published by Allen & Unwin.

Home-made composts

Less precise but quite satisfactory composts, based on the J.I. formulas, can be made by using a good medium garden soil (not a chalky one), a high-grade granulated horticultural peat and coarse sand or grit.

Before use the soil should be sterilised. Although small proprietary sterilisers are available, it is often possible to make do with no more than a 6pt saucepan. Half a pint of water is brought to the boil in this and the pan is then filled with dust-dry soil which has been put through a ¼in sieve. Boiling is then continued for 7min, after which the pan is allowed to stand for a further 7min before the soil is poured out on to a clean surface and allowed to cool.

Neither the peat nor the sand needs sterilising, but the former should be put through a ½in sieve before the composts are made up as follows.

Seed compost: 2 parts soil, 1 part each of peat and sand. To each 8gal of the mixture add ¾oz of chalk (carbonate of lime or ground limestone) and 1½oz of superphosphate of lime.

Potting compost: 7 parts soil, 3 of peat, 2 of sand. To each 8gal add $\frac{3}{4}$oz of chalk and 4oz of the John Innes Base Fertiliser (see below). For more durable composts the amounts of chalk and fertiliser can be doubled or trebled.

The J.I. Base Fertiliser: this is a complete fertiliser supplying nitrogen, phosphorus and potash. It keeps well and is suitable for general use in the garden and greenhouse as well as for the composts. For those who wish to make it up themselves (although it is much easier to buy it ready-made), the recipe is (parts by weight): 2 parts hoof and horn, $\frac{1}{8}$in grist, 13 per cent nitrogen; 2 parts superphosphate of lime (18 per cent phosphoric acid) and 1 part sulphate of potash (48 per cent potash).

Thorough mixing of the compost ingredients and the fertiliser is important. The bulk materials should be mixed thoroughly first by building them up in layers in a flat heap and turning them several times. The fertilisers should then be mixed with a little dry sand and sprinkled gradually over the heap, which should be turned at the same time until there are no unmixed areas.

Soilless composts

Although rather expensive, these offer a convenient and simple way of getting good results. They have become increasingly popular for both seed-sowing and potting and are readily available at garden shops, usually in the form of a peat or peat and sand mixture plus fertilisers. (Owing to the accuracy with which they have to be prepared, it is seldom satisfactory to make them up at home.)

OTHER REQUIREMENTS

A good supply of water is obviously essential and if possible this should be provided by means of a tap in the greenhouse connected to the main supply. A large tub filled by a hose-pipe is the next best thing, but the tub should be covered as otherwise the action of light will soon turn the water green. A fairly large watering-can with a fine rose will come in useful for watering boxes of seedlings and plants in large pots, but a much smaller one (or even an old teapot or kettle) enables the watering of most pot plants to be done quickly and efficiently, particularly if there are pot saucers with the plants.

These pot saucers are available in the same diameters as plant pots and in my experience they provide the most satisfactory way of watering plants in pots. They do at least ensure that the lower roots get their fair share, which does not always happen when the plants are watered from above.

Among the other items needed are a good syringe for use with insecticides, a supply of labels and some small canes and soft green twine for the staking of those plants which need it. A dibber about the size of a pencil will also be needed for the transplanting of small seedlings and the insertion of cuttings. Another useful item is a potting stick, which is simply a piece of wood about 15in long and 1in thick (a piece of broomstick, slightly tapered at one end, is ideal) used for firming the soil well down into the bottom of large pots when repotting is being done.

Lastly, one or two wooden 'pressers' are essential for the firming and levelling of composts in pots and boxes. These are no more than rectangular or circular pieces of wood about 1in thick, with another piece attached to form a handle (see Fig 1).

The cold frame and plunge bed

A cold frame can not only save a good deal of space in the greenhouse but is also essential for the hardening off of summer bedding plants to get them used to the outside weather conditions. A well-drained position in full sun is the most suitable, and to be of really practical use the frame should be of a good size (most of those offered for sale are too small). Apart from holding enough bedding plants to relieve the pressure on greenhouse space in spring, it can also be used for those pot plants that are being grown through the summer to flower from autumn onwards.

For this latter purpose the frame is best made in the form of a plunge bed. In this the plants can be buried to the pot rim so that they stay evenly moist without much watering. Nothing very elaborate is involved in this. All that one needs to do is to excavate the soil in the frame to a depth of about 9in, put in a layer of rubble to provide drainage, and then fill up to ground level with a mixture of moist peat and sand. If the ground is not too well drained, a similar bed can be prepared above ground level by making the frame deeper and filling the bottom of it with the peat-sand mixture to a depth of about 6in.

Whether or not it is made as a plunge bed, the frame should be at least 9in high at the front rising to say, 15in at the back. This will allow for the height of fairly tall plants while the lower-growing ones, such as summer bedding plants in boxes, can easily be kept nearer the light by standing them on bricks or pots.

Chapter 3

Succeeding with Seeds

To be able to produce constant and reliable results with seed is one of the most important aspects of greenhouse work and contrary to many amateurs' experience there should be nothing 'hit or miss' about it.

The first essential in getting a good germination lies in the quality of the seed itself. There is little need to worry about this, as most of the seed sold these days is of such high standard that failures due to poor viability are comparatively rare. The cause of any failure must therefore be looked for elsewhere and in this connection it is important to appreciate the main requirements for successful germination: warmth, air and moisture.

Naturally, the amount of warmth needed depends on the sort of seed, and information on this point is usually readily available either on the seed packet or in gardening literature. To exceed this amount of warmth will in most cases do no harm provided the seeds are kept adequately moist, but to attempt to germinate seeds in temperatures much lower than those recommended is merely asking for trouble.

This need for adequate warmth is generally recognised by the amateur but less so is the equally important need for air, without which the seeds will not germinate and grow satisfactorily. Usually the two main causes of lack of air are a badly drained, waterlogged soil and too deep sowing, so that it is most important to use a sufficiently open and well-drained compost and to sow the seeds at the correct depth, which depends upon their size. Very tiny seeds—and there are some that are no larger than fine dust—are best sown merely on the surface of the compost, while larger ones should be covered to a depth equal to no more than about twice their diameter. It is better to sow too near the surface than too deeply, for in suitable conditions many seeds will germinate with no covering at all; if this were not the case, we would not be troubled with so many weeds as we are!

Perhaps the biggest problem for the amateur is the amount of moisture that the seeds need, the most common fault being to let them get too dry. In most gardening literature the dangers of over-watering are usually so emphasised that the poor beginner is almost scared of keeping his plants wet at all and yet with seed far more failures are due to dryness than to over-watering. Given a well-drained, open compost and sufficient warmth there is in fact little danger of over-watering seeds but even a very short period of dryness may have fatal results, particularly if it occurs after the seed has just started into growth, when as a result of too little water the tiny emerging shoot dies and the seed becomes in effect 'malted'.

Sowing in seed trays

The containers to be used must be perfectly clean and although it is sometimes recommended that the drainage holes should be covered with crocks, this is quite unnecessary for seed-sowing. If a layer of rough peat is placed over them instead this will be quite sufficient and it will also help to economise on the amount of compost needed.

When filling the seed tray with a seed compost of the John Innes type, this must be made evenly firm throughout by working it well down with the fingers into the sides and corners and finishing off with the presser. If a soilless compost is used, firming down is not so important: with this type of compost it is possible to make it *too* compact, so that with continued watering it eventually becomes a soggy, waterlogged mass. As long as a compost of this sort has no really loose areas, it will be firm enough.

A correctly filled seed pan or tray should have a smooth level surface finishing about ½in below the rim, or ¼in in the case of extremely small seeds. These last are the most difficult to sow, but the job will be made easier if the surface of the compost is first covered with silver sand (obtainable at garden shops) and the seed tray is raised to eye-level so that it is possible to look across the surface rather than down upon it. The seed packet is then formed into a spout and held almost horizontally while it is shaken gently, when it should be possible to see the seeds as they fall and thus to get a fairly even spread.

These tiny seeds need no covering of compost but they should be lightly firmed in with the presser. Make sure that this is dry first, otherwise it may pick up the seeds. Larger seeds, which should be covered as mentioned earlier, should have the covering pressed down gently over them. In some cases, notably with tomatoes and cyclamen, these larger seeds are best sown individually about an inch apart, and although this is a more troublesome method it means that there is not the same urgency to prick them out as there would otherwise be.

Watering seeds

After the seed has been sown, the container should be stood almost up to its rim in water until the moisture seeps through to the surface, which should be allowed to become quite wet although not to such an extent that the seed can float about. Subsequent waterings should then be carried out in the same way while the surface is still moist, or in the case of the large seeds that have been covered with compost a watering-can with a fine rose may be used instead.

Covering the seed tray with glass and paper will help to reduce the amount of watering needed and will also assist germination, as most seeds germinate best in the dark. But at the first sign of the seedlings, these coverings must be removed and the plants grown on in full light, with shade only from full sun.

The propagating frame

The use of some sort of propagator is desirable but not actually essential. Although it enables the seeds to be sown earlier than might otherwise be possible this is of little use if the greenhouse itself cannot be maintained at the temperature necessary to cope with the seedlings. In any case many of the seeds commonly sown in greenhouses will germinate satisfactorily on the open bench provided a minimum temperature of 50–55° F (10–13° C) can be kept up, and although they may do so more slowly than in a propagator, they will at least be acclimatised to greenhouse conditions as soon as they appear.

Seeds requiring higher temperatures than this must be raised in a propagator or in a warm room indoors, but even then it must be borne in mind that in two or three weeks' time they will have to be transferred to adequate warmth in the greenhouse. If this is not possible, it is best to make the sowings rather later, when warmer conditions can be maintained more easily. This applies particularly if the greenhouse is not heated at all, in which case there are very few sowings that should be made before mid-March at the earliest.

Chapter 4

Taking Cuttings

Next to seeds, cuttings of one sort or another provide the main way of raising a stock of plants and with some subjects this is the only method in common use. For instance, many plants such as chrysanthemums, perpetual-flowering carnations and certain types of begonia do not come true from seed—that is, the seedlings may not have exactly the same characteristics as the parents—and for these the vegetative method of propagation by cuttings has to be used where named varieties, or cultivars as these are now known, are to be grown.

Although there are several different types of cutting, as will be described later, the process of root formation on all of these is the same. First, when the cutting is severed from the plant, a protective layer of cork cells forms over the cut end to prevent the loss of moisture. The cells adjacent to this then divide and multiply to form a further layer known as a callus, which on many cuttings can be readily seen as a sort of knobbly growth. Finally, as a result of hormone activity within the plant, roots develop in some of the cells forming the callus and when these eventually break through the surface, the cutting is said to have 'taken' or 'struck'.

To enable these different stages to take place satisfactorily, certain conditions are essential, and these must be provided by the compost and the atmosphere in which the cuttings are being rooted. An adequate supply of air at the base of the cuttings is one of the most important conditions and the main ingredient of the compost is therefore sand, grit or vermiculite: each of these readily permits the passage of air. But moisture is also essential and although rooting will normally take place in the sand or grit alone, provided this is kept wet enough, it is usual except in certain cases to include an equal quantity of granulated peat as well.

This mixture of equal parts peat and sand will suit most cuttings, but for some strong growers like chrysanthemums and dahlias a richer mixture such as the J.I. Seed Compost, or even the J.I.P1, can be used. On the other hand, some cuttings, notably those of the indoor plant Busy Lizzie, *Impatiens sultanii*, will root readily enough in water alone: this is a method that is well worth trying for other plants, since it involves no more than taking a few extra ones and putting these in a jar of water in a warm room.

In the case of cuttings which are taken complete with their leaves—and these include the very common type of 'softwood' shoot cutting which will be dealt with later—it is also essential to provide a humid atmosphere. Without this the leaves would soon give off or 'transpire' the moisture already in the cutting, which having no roots to take up more, would soon shrivel and die. In a humid atmosphere, however, the leaves become covered with a film of moisture which prevents or at least hinders the loss of moisture. In this way the cutting is kept fresh and turgid until it has made some roots and is thus able to look after itself.

The close conditions inside a propagating frame, or better still in a mist unit, are ideal for providing this humid atmosphere; but even where the cuttings are to be rooted on the open greenhouse bench—a common practice with chrysanthemum cuttings—it is usually possible to provide the necessary humidity by shading them from the sun, standing them on a moist base and syringing them frequently overhead.

A certain amount of heat is usually necessary too, although in summer this need be no more than the normal temperature, without any artificial aid. It is a different matter in winter and early spring, when a temperature of 40–45° F (4–7° C) is the very minimum that will be needed, with one a good deal higher for the cuttings of most plants. As far as possible this heat should be applied from below ('bottom heat'), as here it stimulates root rather than top growth. This is one good reason for positioning the propagating frame over part of the heating system or, alternatively, for using a soil-heating unit.

SHOOT CUTTINGS

The most common type of cutting consists of the terminal portion of a stem or sideshoot. Cuttings of this type are divided largely into two sorts, 'hardwood' and 'softwood'. The former are seldom used in greenhouse work, but the latter are very commonly used and provide a ready means of increasing most plants that form definite stems or branches.

The term 'softwood' is used because cuttings of this type are taken while the shoot is still soft and green; in some cases it may be just changing from this green, sappy state into more mature and harder wood, when it usually takes on a darker or reddish appearance. At this stage it is said to be 'half-ripe'. Cuttings of this sort are commonly used for the propagation of the zonal pelargonium (better known as the geranium), which owing to the succulent nature of the shoots seldom succeeds if the cuttings are taken in the fully green state.

For most plants cuttings 2–3in long are suitable, preference being given to the smaller size, since for some reason or other the smaller the cutting the quicker its roots, provided it is large enough to be anchored securely in the rooting medium. Unflowered shoots, neither too thin and hard nor too thick and soft, and with the leaves set closely together (short-jointed) are the ideal ones to use. In most cases they should be trimmed off immediately beneath a leaf-joint, using a sharp knife or pair of scissors, after first removing the lower leaves. Cuttings trimmed off in this way are said to be nodal cuttings, but in some cases it is better to trim them off in between the leaf-joints or nodes, in which case the cutting becomes an internodal one.

Shoots suitable for cuttings are produced in different ways on different plants. On some plants such as the fuchsia and the zonal pelargonium, shoots of the current year's growth may be taken off during the summer or the old plants may be cut back in spring to encourage the production on the old stems of suitable shoots for cuttings. This latter method is practised on many plants but in the case of the chrysanthemum, for instance, the plant is cut back in a similar way but with the idea of producing

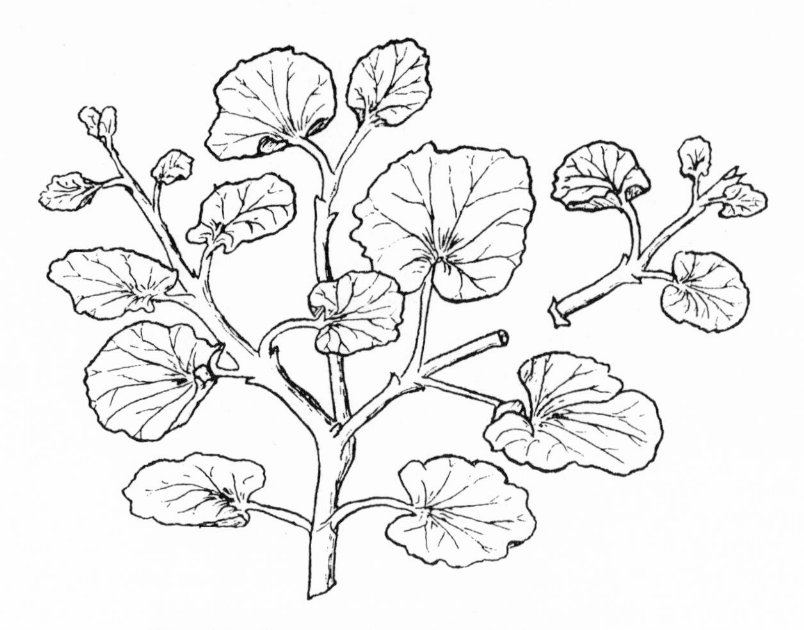

Fig 2(a) Shoot cutting of zonal pelargonium

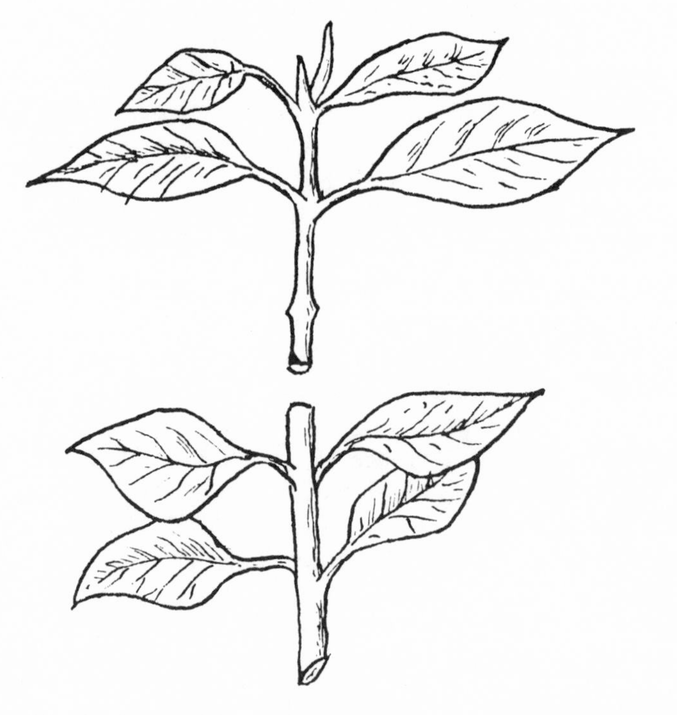

Fig 2(b) An internodal fuchsia cutting, with the cut made between the leaf-joints

Fig 2(c) A basal cutting (chrysanthemum)

shoots from ground level (basal shoots). Some tuberous plants, such as the dahlia, are again treated differently by starting the old tubers into growth in spring and using as cuttings the young shoots which spring from them.

Inserting the cuttings

This presents no problem as it is necessary only to dibble them firmly into the rooting medium, just close enough to allow a circulation of air around them. Small quantities can be inserted round the edge of a plant pot, while larger quantities may go either into seed trays or into a bed of suitable material in the propagating frame. The main thing is to make sure that the cuttings are quite firmly in, with the base of each one in contact with the rooting medium, but this does not mean that they should be inserted deeply. The nearer the base of the cutting is to the surface, the quicker it will send out roots. Cuttings should therefore be put in just deep enough to keep them erect after they have been given a thorough watering; they will then be ready to go into the propagating frame or on to the greenhouse bench, where they should be given only enough water to keep them moist.

LEAF-BUD CUTTINGS

The stems of some plants, particularly evergreens such as *Camellia iaponica*, *Hederas* (Ivies) and *Ficus elastica* (the Indiarubber plant) can also be used to provide leaf-bud cuttings.

All one needs to do is to cut a half-ripe stem into short portions each containing a mature healthy leaf. (There are one or two variations on this, such as scooping out the leaf with a shield-shaped piece of the stem attached or trimming off the stem on the side away from the leaf so that a flat area of tissue is exposed.) Most leaf-bud cuttings will, however, root without any special

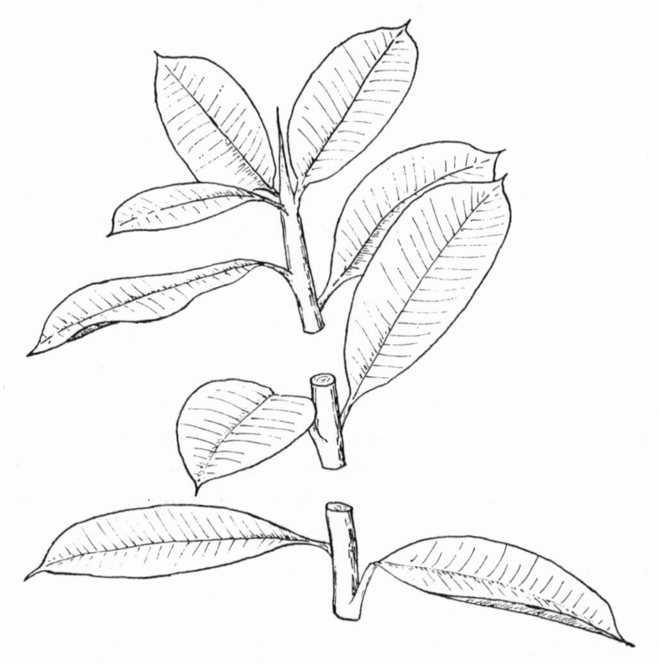

Fig 2(d) *Ficus elastica*, showing a terminal shoot cutting at the top and a leaf-bud cutting at the centre

treatment provided there is a dormant bud at the junction of the leaf and the stem, for this is the bud from which the new shoot will grow.

Cuttings of this type are inserted and treated in exactly the same way as shoot cuttings, but with the dormant bud just beneath the surface.

LEAF-STALK CUTTINGS

Some plants, mainly the members of the Gesneriaceae family which includes such popular subjects as gloxinias, Saintpaulias (African Violets) and streptocarpus, do not make any shoot growth and they are therefore propagated by leaf-stalk or leaf cuttings. For leaf-stalk cuttings, healthy mature leaves are taken with 1–1½in of their stalks and are inserted and treated in much the same way as shoot cuttings. They should be set upright in the rooting medium so that the stalk is buried up to the leaf-blade. Young plantlets will then form at the base of the leaf-blade and as soon as these are large enough they are taken off and potted up.

For easier handling and also to reduce the loss of moisture through transpiration from the leaves, it is usual to cut the top half off large leaves, such as those of the streptocarpus, before inserting the cuttings.

LEAF CUTTINGS

This method, which uses the leaf-blade only, is most commonly followed for gloxinias and the foliage plant *Begonia rex*. After a healthy mature leaf has been removed, its main veins are slit across with a sharp knife and the leaf is then laid face upwards on a moist rooting medium, where it should preferably be pegged down with fine wire to keep it in position. In warm humid conditions young plants should form at the slits in from six to eight weeks, and when these are potted up the remaining portion of old leaf is trimmed away.

In commercial practice a more rapid increase of *Begonia rex* is obtained by cutting the leaf into 1in squares, which are merely laid on the surface of the rooting medium. Or another way still is to use triangular portions of a similar size, each containing a junction of the main veins at one corner of the triangle. These leaf portions are then inserted upright with the junction of the veins being buried.

LAYERS

Although used mostly for outdoor plants, layers can be taken from some greenhouse climbing plants. The simplest way is to peg down one of the stems, at a leaf joint, into a small pot of soil; but quicker results are obtained by making a 'tongue' first. This is done by cutting halfway through the stem immediately below a joint and continuing the cut upwards towards the end of the shoot for about an inch. The two halves of stem so formed are then pegged down into the pot of soil, keeping them apart.

AIR-LAYERING

This is most commonly carried out on the Indiarubber Plant,

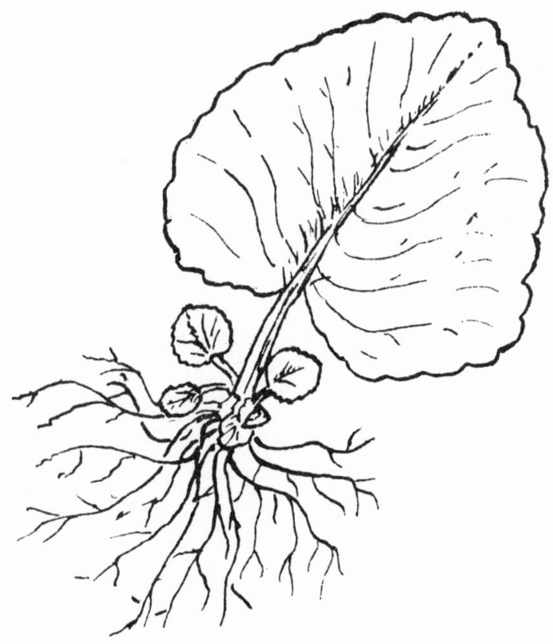

Fig 2(e) A rooted Saintpaulia leaf-stalk cutting, with new plantlets at the base

Page 33 The ideal in greenhouse ventilation: a continuous run of opening lights at both ridge and sides

Page 34 (*above*) Slatted roller blinds are excellent for providing summer shade; (*below*) staging must be strongly supported to take the considerable weight of the plants

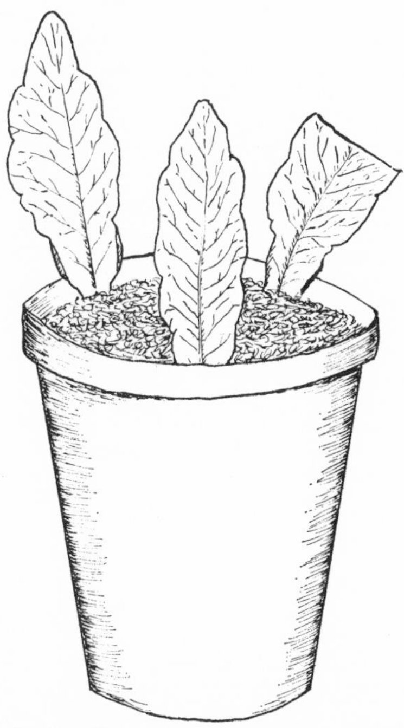

Fig 2(f) Gloxinia leaf-stalk cuttings inserted round the edge of a pot. Where height in the propagating frame is limited the upper half of the leaf may be cut off, as at right

Ficus elastica, and it provides an easy way of raising a new plant from one that has become old and 'leggy'.

The upper leaves are first tied together to keep them out of the way and the leaves immediately below them are then removed to leave the stem bare for 3in or 4in. The next thing is to make a slanting cut upwards, starting just below a joint and continuing the cut upwards and inwards for about 1½in into the bared stem. Damp sphagnum moss, obtainable from garden shops or florists, is next packed round the cut and worked in between the two cut surfaces until it forms a thick wad round the stem. This in turn is covered with polythene sheeting, tied round the stem above and below the moss. The plant is then kept in a warm place until roots can be seen in the moss, when the rooted upper portion of the plant is cut off and potted up.

Chapter 5

Pricking Out and Potting Up

Soon after seedlings or rooted cuttings start to grow they will need more light, space and food and in the case of seedlings this is provided by pricking them out or transplanting them into seed trays.

Pricking out

This is best done at a quite early stage, for if the seedlings are left too long in the seed pans they will not only become weak and 'drawn' as a result of overcrowding, particularly if germination has been good, but will also be much more difficult to transplant safely. Ideally, the job should be carried out when they are little more than ¼in high, as at this size there is less risk of damage to the roots than when these have become larger and entangled. Where fairly large seeds have been sown individually, however, they can be left to grow to an inch or so high before pricking them out into seed trays or, in some cases, straight into small pots.

The day before the job is started, the seedlings should be given a good watering to make sure that the compost is moist but not wet when they are taken out of it. At the same time the pricking-out trays or boxes should be filled with compost, in the same way as when the original seed trays were prepared for the seed. For most plants the J.I.P1 or a soilless potting compost is suitable. If the former is at all dry, it should be watered and left until it is just moist enough to stay in position when a hole is made with the dibber. Soilless composts will probably be moist enough already when they are placed in the trays. If any watering is needed at all it should not be overdone, otherwise the compost will be too soggy for easy pricking out.

When the actual pricking out is started, a few seedlings should be lifted carefully from the seed pan with the tip of a small trowel or with an ordinary dinner fork. Take care not to damage the roots more than you can help, and do not take out too many plants at once, particularly in hot sunny weather, when the seedlings should be protected from the sun once they have been taken out of the seed compost. It is always advisable, too, to handle them by their leaves rather than by the stem, as these are less easily damaged.

For most plants a suitable spacing in the pricking-out boxes is 1½in each way, which gives fifty-four plants to the standard seed tray. A start is usually made at one corner by making a hole with the dibber and placing the seedling in position with its roots dangling in the hole; then if the dibber is inserted again at an angle, the compost can be levered firmly up against the roots. For seedlings of a rosette type of growth, with no obvious stem, the hole should be made just deep enough to take the roots, so that the plant finishes just 'sitting' on the surface. Those with an obvious stem, on the other hand, should be buried almost up to the seed leaves (these are the first leaves to appear and they may be quite different from the normal type of leaf on the plant).

When all the seedlings are in, give them a gentle but thorough watering with a fine-rosed can, then put them on the bench in the greenhouse, where the compost should be allowed to dry a little to encourage root action. As soon as growth starts, however, watering can be increased gradually; and eventually, when the leaves are almost touching, the plants will be ready for either planting out or potting up.

POTTING UP

Most seedlings, other than those of summer bedding plants, will need to be moved on from the pricked-out boxes into pots, and this also applies to the cuttings that have been rooted. For this first potting, 3in pots are generally used for those plants which will eventually go into 5in ones—the most popular size for the general run of pot plants—but 3½in pots should be used if the plants are to finish in larger ones still. If only small plants are required, this 3½in size is quite large enough for many different types of plant to be grown to maturity.

As for pricking out, the plants should be quite moist at the root before the job is started, and if 3½in pots are to be used these should be crocked with one piece of broken pot placed over the drainage hole (this is not necessary with the 3in size). Needless to say, the pots should be clean.

Have a heap of compost at one end of the bench and a supply of pots ready crocked if necessary at the other end, where a small batch of plants taken from the pricked-out box should also be placed. A pot is then picked up with one hand and half-filled with a handful of compost from the other. The next thing is to place a plant in position in the centre of the pot and at the right level, while further compost is put round it and settled down by giving the pot one or two taps on the bench. Finally, the compost is made firmer still by encircling the stem of the plant with the thumb of one hand and the thumb and forefinger of the other and pressing the compost gently down round the roots. With 3in and 3½in pots, the compost should finish about ¼in and ½in respectively down from the rim.

POTTING ON

This term is used for the operation of moving a plant on from one pot to another of larger size, the most usual moves being from 3in to 5in and 3½in to 6in, although in the case of large plants such as chrysanthemums further shifts still will be needed, into 8in, 9in or 10in pots. In all cases the pots to be used should be well crocked, with one large piece over the drainage hole and several smaller pieces over this.

The first thing, obviously, is to remove the plant from the pot it is already in. After making sure that the soil is moist, this is done by placing one hand over the top of the pot, with the plant stem between the fingers, inverting the pot and giving its rim a sharp tap on the edge of the bench. The plant, with its roots and soil intact (the 'soil-ball'), will then drop into your hand.

The actual potting is done by putting compost in the pot to a quantity that when the plant is standing on it the top of the soil-ball comes to the right level below the rim. For 5in pots this will be about ½in, for 6in about ¾in and for 8in and larger about 2in. Further compost should then be added and firmed down gently with the fingers until the soil-ball is just covered.

For most plants the compost should be made only moderately firm, particularly if it is a soilless one. For the final potting of hardwooded plants and stronger growers such as chrysanthemums, however, it should be made very firm with the help of a potting stick: this is used for ramming down the compost as it is placed in the pot.

Chapter 6

Greenhouse Management

Although this chapter deals mainly with the management of the greenhouse, it can be applied largely to indoor cultivation as well, the only difference being that in a greenhouse it is usually easier to provide the necessary conditions of light, humidity, ventilation and so on. However, with a little care, most of these can be fairly easily coped with indoors; and as far as one of the main factors—correct watering—is concerned, there is little difference wherever the plants are grown.

WATERING

It is probably safe to say that most failures with pot-grown plants are due to faulty watering, so that, conversely, once the art of watering has been mastered there should be little difficulty in growing most plants at least reasonably well. It is impossible to lay down hard and fast rules about watering, as growing conditions can vary so much and in any case different plants have different needs; but it is possible to give a rough guide as how it should be done.

Perhaps the trickiest time of all for watering is immediately after a plant has been potted, for until it has made some new roots it will not be able to cope with a constantly wet soil. The best procedure at this stage is to give the plant a thorough watering, either by soaking it from underneath or by filling the pot to the rim two or three times, and then leave it to become almost dry. A further soaking should see the plant well established in the pot, when watering can be increased gradually.

If the plant is being grown on through the summer there will be little risk of over-watering it, for most plants, when in full growth in a warm atmosphere, will take liberal amounts. If anything, the danger during this summer season is under- rather than over-watering, with the main source of trouble being the drying out of the lower roots. If the soil is dry and the pot is filled to the rim with water, little more than the top inch or so will be wetted, although the surface soil will look quite wet. If 'top watering' of this kind is carried out, therefore, it may be necessary to repeat it two or three times before the soil is soaked right through. A much better way is to water the plants from underneath, using pot saucers. If a plant gets really dry, so much so that it flags badly, the best remedy is to submerge the pot in a bucket of water in the shade and leave it until air bubbles have ceased to rise.

One problem that often besets the beginner is how to know when the soil is dry enough to need watering. In most cases it is possible to tell by the appearance of the surface of the soil, which looks dry and hard when watering is needed. Or if you are unsure about it even then, it can be tested by scratching it with your finger: if it is dry the soil will crumble readily, whereas if it is still moist it will have a muddier feel. Another way is to feel the weight of the pot, which will be noticeably light if the soil is dry. You have only to test the weight of a pot of dry soil against a wet one to see what a great difference there is. Lastly, you can tell by the condition of the leaves, which will be somewhat limp and flaccid if the plant is dry. This is not too reliable a method, however, as strong light after a period of dull weather can produce a similar effect even though the soil is moist.

The most important time to know whether or not a plant is dry is in the winter and early spring, as this is when it is most easy to over-water. For most plants the safest method is to let them almost dry out between thorough soakings, and if you should be in doubt about watering, a good rule is 'Don't'. Few plants will suffer much if they happen to be extra dry for a time in winter as long as they are thoroughly soaked afterwards.

So far I have been dealing with the watering of plants in pots. Much the same principles apply to seedlings and cuttings, except that while they are still in the seed pan the former should never be allowed to dry out completely, as their tiny roots may then soon shrivel and die. With most cuttings, on the other hand, it is better to err on the dry rather than the wet side, as even if they flag a little from dryness, they will soon perk up again after a thorough watering.

FEEDING

Although the food in the potting compost is sufficient to keep the plants going for a period, additional feeding is usually necessary by the time the roots have almost filled the pot, or even before then. Dry fertilisers, sprinkled on the surface and watered in, are sometimes used but liquid ones are more common and there are more than enough of these in proprietary form. There are too the John Innes feeds, which are particularly suitable for use with the J.I. composts. They are available in both dry and liquid forms.

Most of the fertilisers offered for use in the greenhouse or home are 'complete' ones in that they contain the three main plant nutrients, nitrogen, phosphate and potash; the proportion of each will vary according to the type of fertiliser. These proportions are given in percentage form on the container, with nitrogen indicated as N, phosphate (as phosphoric acid) as P_2O_5 and potash as K_2O.

It is thus easy to see in which of the three nutrients the fertiliser is particularly rich, but this is of little use unless you know what effect each one has.

Nitrogen produces strong vegetative growth, resulting in large lush leaves. It is useful therefore for plants which are making poor stunted growth, with small and badly coloured foliage. If used in excess, however, it produces soft sappy plants that are not only very susceptible to disease but which also will probably produce very poor-quality flowers and fruit—or even none at all!

Potash has more or less the reverse effect of nitrogen in that it helps to produce firm healthy growth and plants, flowers and fruit of good quality; in fact its effect is something like that of sunlight. In addition to its inclusion in complete fertilisers it is therefore sometimes used on its own, as sulphate of potash, to remedy the effects of too much nitrogen or even of dull damp weather, particularly in tomato-growing. Too much potash may, however, lead to very hard slow growth, with delayed flower and fruit production; too little is generally indicated by a scorched mottled appearance of the leaves, which do not then function properly.

The effect of phosphate is less obvious but in the seedling stage it helps to produce good root and top growth. This is why it is generally included in seed composts. To a lesser degree it also hardens the plant against disease and low temperatures and helps to mature it to the fruiting stage, but used in excess it carries this process too far and produces hard stunted growth.

It is better to start feeding too early rather than too late, and if a start is made when the leaves have spread out to the side of the pot, it will not usually be too soon. It is, however, no use trying to feed a plant with poor roots and in such a case the best thing to do is to encourage root growth by keeping the plant a little on the dry side. Incidentally, the roots of most pot plants can be examined by knocking them out of the pots, when it should be possible to see plenty of healthy white roots. If the roots are brown and shrivelled, usually something is wrong. It is necessary too to ensure that the soil is moist before either liquid or dry feeds are applied, as otherwise these may damage the roots. Lastly, if any fertiliser should get on to the foliage, wash it off immediately using the watering-can.

VENTILATION, SHADING AND DAMPING DOWN

Like people, plants do best when they get plenty of fresh air and correct ventilation is therefore an important part of greenhouse management. In summer this usually presents no problem, as it is safe to ventilate quite freely then, even to the extent of leaving the greenhouse door open day and night in hot weather; but at other times of the year, ventilation can be a tricky business in our changeable climate.

In hot weather the purpose of the ventilators is obviously to keep the temperature down by allowing the warm air to escape, but when artificial heat has to be used, the problem arises of conserving this as much as possible while keeping the air fresh. This problem is usually overcome by leaving the ventilators open an inch or so on the side away from the wind during the night, and then gradually admitting more air during the day as the weather permits. The only time the ventilators should be closed completely is during foggy weather. At all other times they should be kept open sufficiently to create a good circulation of air, without draughts.

The cool damp days of autumn are perhaps the most difficult time. Although artificial heat may not be needed then, the air is often so still and moist that it circulates hardly at all in the greenhouse. At such times it is usually wisest to put on slight heat if only to dry the air a little and keep it on the move. Without this the dampness may settle on the blooms of plants, particularly chrysanthemums, and cause serious damage. Even worse, the plants may be attacked by one of the more common fungus diseases such as grey mould (see Chapter 7).

Damping down

In hot weather the greenhouse temperature can be lowered by damping down, which consists merely of making the floor, walls and staging thoroughly damp. In small greenhouses this is easily done with the watering-can, and the effect will be immediately apparent, not only in the drop in temperature but also in the atmosphere, which becomes noticeably fresh and damp. It is this humidity which creates the 'growing atmosphere' that one expects in a greenhouse.

Damping down may be continued safely from the first hot days of spring until the autumn, but then must cease as it becomes essential to keep the greenhouse dry enough rather than damp.

Shading

This provides another way of keeping the temperature down in summer, but it must not be overdone or the plants will grow tall and weak. Apart from some ferns and the foliage plant *Begonia rex*, very few plants are happy in cave-like conditions and all that should be aimed at therefore is filtered shade, just enough to break the full force of the sun. For this purpose there are proprietary compounds which are made into solutions for spraying on to the glass; but a similar effect can be obtained equally well with a thin paste made of flour and water which has the advantage of being translucent. The trouble with these semi-permanent shadings, however, is that they cannot be removed easily when the weather changes. A much better, if more expensive, method is to use roller blinds. It is possible to buy these made of wooden slats, but simpler ones can be fashioned easily at home by using green plastic film or even muslin attached to rollers, so that the blinds can be put up and down by hand.

For very small greenhouses another useful method of shading is to make some light wooden frames and to cover these with thin wooden laths about an inch wide and spaced an inch apart. The frames can then be laid on the greenhouse roof or against the sides and taken down again when the sun has gone in. Similar light frames covered with green plastic film or muslin may be used also inside the greenhouse.

'STOPPING' AND STAKING

'Stopping' or pinching out is done with the object of making a plant branch out earlier than it would have done. Carried out while the terminal portions of the stems are still soft and green, it involves no more than the removal of the topmost point, sometimes only back to the topmost pair of leaves, although in some cases the stem is shortened well back. In either instance the portion removed should be severed just above a leaf-joint, as the wound will heal more rapidly here than if a stump is left above the joint.

Only plants of branching habit need to be stopped, and not all of these. Except in the case of chrysanthemums, where the number of flowers and the time they appear are often regulated by stopping, it often pays to grow similar plants with and without stopping, to see which plant gives the best results. Certain varieties of fuchsia, for instance, branch out readily on their own while others will not do so without stopping.

Staking

A good many plants will also need staking to keep them neat and upright and this should be done long before the stems tend to flop over. Small green-painted split canes can usually be purchased at garden shops and these are ideal for the job, particularly if they are used with soft green garden twine, which is much easier to handle than the more old-fashioned raffia. The art of staking consists of making the supports as inconspicuous as possible and the more neatly it is done the less it will detract from the appearance of the plant.

CLEANLINESS

Apart from looking far more attractive than an untidy one, a clean and well-ordered greenhouse usually means less in the way of trouble with pests and diseases, for the source of these is often in decaying or dead plant material left lying about or beneath piles of dirty boxes and pots. A constant clearing out of weeds and fallen leaves, faded flowers and finished plants is necessary. Overcrowding of the plants should be avoided too as far as possible, since this makes both work and the control of pests and diseases more difficult.

THE COLD FRAME

Much that has been said in connection with the greenhouse applies also to the cold frame, which is even more apt to become full of weeds and other debris, at any rate during the period when it is not in use. The remedy, obviously, is to keep the frame in use as much as possible.

As with the greenhouse, ventilation and shading are important: whenever the frame has been closed completely at night, it should be opened first thing in the morning, even if only a little in bad weather. This will be sufficient to freshen up the air inside it. In fine weather it will need to be opened both earlier and wider, otherwise, as soon as the sun reaches it, the temperature inside will go up rapidly, with possible scorching of the plants. At the end of the day it should be closed just before the sun goes down, so that the heat of the day is retained in it, although in warm summer weather there is seldom any need to close it completely. In fact, in summer, the glazed frame-light can often be left off altogether and replaced with a slatted frame to create shady but airy conditions, the glazed light being put back only if there is a likelihood of heavy rain.

Hardening off

One of the main uses of the cold frame is for the hardening off of summer bedding plants which, having been grown in the warm conditions of the greenhouse, are far too soft and tender to be put straight outside. However, as soon as there is no risk of severe frost, the earliest of them can go in the frame, where they must be protected from any check due to cold nights and biting winds. For this purpose the frame should be closed completely for the first few nights and opened only slightly during the day unless the weather is very mild. More ventilation can then be given gradually both night and day, until eventually the plants are open to the air all day and finally all night. But always be prepared to close the frame and even cover it with sacks or matting if there happens to be a late frost, and do not forget that cold winds can do almost as much damage as a frost.

Chapter 7

Pests and Diseases

Even in a well-kept greenhouse, pests and diseases are sure to put in an appearance sooner or later. There are fortunately very few that cannot be controlled without much difficulty at an early stage, but if they are left to multiply or spread it may become almost impossible to deal with them.

Greenfly

This common name covers a number of related insects, known as aphides. Being sap-sucking insects, their presence is usually first indicated by the unhealthy appearance of the leaves, which may become discoloured and distorted. In addition there may be a covering of a sticky substance, euphemistically called honey dew, which is deposited by them (see also 'Sooty mould' later in this chapter). Where indoor plants are badly infested, incidentally, this honey dew can drip off the plant and spoil polished surfaces, so it becomes all the more important to keep the plants clean.

Usually, spraying the plants with derris, malathion, nicotine or BHC (lindane) will control minor infestations, while for more serious ones fumigation with either nicotine or BHC is more effective.

Glasshouse leafhopper

This is another sap-sucking insect which has much the same effect as greenfly and which deposits honey dew in the same way. In adult stage it is similar to a pale yellow greenfly but is much more active, running from leaf to leaf, so that it is not as readily seen as is the less mobile greenfly. Its presence is, however, easily detected by the white 'moult-skin' that it sheds at a certain stage of its life and which remains attached, more or less in the shape of an insect, to the leaf. The beginner at greenhouse work is often misled by the presence of moult-skins into thinking that his plants have been attacked by another and much more serious pest, the greenhouse whitefly, but this is a quite different problem which will be dealt with later in this chapter.

The glasshouse leafhopper can be controlled in the same way as greenfly, but the most effective answer to it lies in dusting the plants with nicotine dust, obtainable at garden shops.

Mealy bug and scale

Honey dew may also appear on plants infested with another group of sap-sucking insects which fall into two classes: mealy bugs and scale insects. Although in general these are somewhat similar, they can be readily distinguished from each other by the fact that the mealy bugs have a greyish-white waxy covering while the scale insects usually have a hard, brown, shell-like coat. Both are about ⅛in long and oval in shape, but while the mealy bugs are fairly active the scale insects rarely move.

Both these pests attack many greenhouse plants, particularly evergreen ones, but as they are often in the more inaccessible parts of the plant they can escape notice until honey dew puts in an appearance. Owing to their waxen or hard coverings these pests are not easily dealt with but malathion, applied as a spray thoroughly and forcibly at fortnightly intervals, will usually effect a satisfactory control. Small infestations may often be removed by merely scraping them off.

Red spider mite

The damage this pest does is out of all proportion to its size, which is less than a pinhead. It is not a spider at all but belongs to a family of mites, some of which spin webs similar to those of spiders. The web covers the minute eggs and on badly infested plants it can be clearly seen together with a bleached and mottled appearance of the leaves, which may become suffocated by it to such an extent that they wither and fall. In spite of the name 'red spider' only the overwintering females, found in cracks and crevices in brickwork, woodwork, canes and so on, are actually red; for the rest of the year the mites are more or less straw-coloured.

Where this pest appears it is usually an indication that the atmosphere is too hot and dry, for it is in such conditions that the red spider is most troublesome. Prevention, therefore, lies largely in creating greater humidity by shading and damping down, but if a plant does become infested it should be put outside and sprayed repeatedly with nicotine or malathion. For more severe infestations the only satisfactory control is fumigation with azobenzine, with a second dose about ten days later.

Greenhouse whitefly

The effect of this pest, known also as the snowy or ghost fly, is very much the same as that of the other sap-sucking insects, but the insect iself is very different from those already mentioned. Both in flight and at rest it might easily be taken for a minute white moth, less than ⅛in long. If a badly infested plant is shaken, these tiny 'moths' will rise in clouds and flutter about until they settle on other plants. Many types of plant, but particularly tomatoes, fuchsias and salvias, are attacked and it is most importhant to take control measures as soon as the first flies are seen. Otherwise, the whole stock of plants will soon become infested.

The best control undoubtedly lies in the parasitic wasp *Encarsia formosa*, an even more minute creature, which is practically invisible to the naked eye. It lays its eggs in the larval stage of the whitefly; then when these eggs hatch out the grubs feed on the larvae, which eventually turn black as they are destroyed. Finally, when there are no more larvae, the parasite itself disappears.

Unfortunately, supplies of this parasite are not as easy to obtain as they used to be, although it may be possible to get a supply through a horticultural organisation. If not, recourse will have to be made to chemical control, using BHC or malathion as an aerosol or fumigant.

Page 51 (*right*) When sowing or pricking out, a bottom layer of rough peat reduces the amount of compost needed

(*left*) The compost for seed-sowing and pricking out must be well firmed down, particularly at the corners

(*right*) The surface of the compost is made smooth and level with the presser

Page 52 (*left*) The seed is spread as evenly as possible

(*right*) Covering the seed tray with glass and paper prevents the compost from drying out too quickly

(*left*) Pricking out is best done while the seedlings are still quite small

Vine weevil

The damage done by this insidious pest is usually not noticed until it is too late, for often the first indication that a plant has been attacked is its total collapse. Actually it is not the weevil itself (which is something like a black, faintly speckled beetle about $\frac{1}{3}$in long) that does the damage, but its larvae or grubs, which eat away the roots of many greenhouse plants including cyclamen and primulas. During the summer the weevil lays its eggs at the base of the plant, then in winter these hatch out into fat, greyish-white grubs about $\frac{1}{4}$in across when they are in their characteristically curved position. If an attacked plant is shaken out of its soil, you will probably find two or three of these grubs at its roots.

Control lies largely in keeping the weevils at bay during their egg-laying period in summer, when BHC dust should be sprinkled round the plants, with a little also being used to mix with the potting composts. The only thing to do with plants that have collapsed owing to this pest is to repot them in fresh compost and encourage them to make new roots.

Eelworm

This comes in various types attacking begonias, ferns, bulbs and chrysanthemums as well as other plants. In general it is a minute, transparent, worm-like creature, small enough to be invisible to the naked eye. It lives in the tissues of leaves and stems, with the result that they become distorted and disfigured. On chrysanthemums, for instance, the leaves turn brown, then black, and finally hang down and wither.

There is no means of control by insecticides and with chrysanthemums the usual treatment is to immerse the dormant roots for twenty to thirty minutes in water at a constant temperature of 110° F (43–44° C). A similar method is used for bulbs. Unfortunately, this treatment is not easy for the amateur to carry out and the safest way is to discard affected stock and start again with new plants in fresh soil.

Tarnished plant bug

Although it attacks other plants as well, this is a particular nuisance on chrysanthemums, which develop lop-sided or otherwise distorted blooms and damaged and deformed leaves, together with possible 'blind' or flowerless shoots. It is fairly easily recognisable as a yellowish-green, long-legged insect about ¼in long, with a pattern in the form of a bishop's mitre on its back (it is known also as the bishop bug), but owing to its habit of dropping to the ground when disturbed it is not too easy to find or control. Dusting the plants with DDT from July to September is the best method of coping with it, but since this insecticide is somewhat frowned upon now, you may prefer to use the less effective malathion as a spray during the same period. In addition all weeds of the Compositae family, with single or double daisy-like flowers, should be kept down as these act as host plants for the pest.

Chrysanthemum leaf-miner

Despite its name this pest attacks other plants as well, notably cinerarias. Its presence is soon indicated by the pale, meandering tracks it makes as it burrows between the two outer surfaces of the leaf. Immediately these tracks are seen, the plants should be sprayed thoroughly with BHC, malathion or nicotine.

Slugs

These are often an unsuspected source of trouble and if seedlings, leaves, bulbs or other plant material are being eaten it is just as well to look around for these pests, whose presence is sometimes indicated by the trail of slime which they leave. They are often brought into the greenhouse on the bottom of boxes and pots. Fortunately, they are easily dealt with these days by using one of the many proprietary slug-poison baits, which should be put down as soon as the trouble is suspected.

DISEASES

Damping off

Until the introduction of seed composts with sterilised or no soil in them, this was a very common trouble with seedlings, and it can still be a serious nuisance where cultural conditions are not all they might be. It is caused by a soil-borne fungus that attacks the base of the stem, with the result that this shrivels and dies. The whole plant collapses even though the rest of it may look quite healthy.

If the trouble does start, the seedlings should be pricked out immediately, with care taken not to include any that show signs of discoloration or narrowing at the base of the stem. The pricked-out boxes should then be watered with Cheshnut Compound, which can be readily obtained at garden shops. It is also advisable to water the seedlings with this before they are pricked out, or as soon as the trouble has been noticed.

Persistent trouble with damping off usually means there is something wrong with the growing conditions. Too little warmth in particular will soon lead to trouble—damping off seldom occurs in the warm airy conditions of summer—but too little ventilation and faulty watering will also encourage the growth and spread of the fungus.

Foot rot

Known also by other names such as collar rot and black neck, this is in effect a form of damping off that attacks cuttings and older plants. On these the base of the stem decays. It is generally caused by over-watering combined with lack of warmth and ventilation. As there is no cure, affected plants should be destroyed immediately.

Powdery mildew

This fungus, typified by the white floury coating often found on roses out-of-doors in summer, is more prevalent in the garden than in the greenhouse but nevertheless it commonly attacks some greenhouse plants, mainly grape-vines and chrysanthemums.

Prevention and control consist largely of providing a free and ample circulation of air around the plants, making sure that they are never too dry at the roots and spraying them with a fungicide based on copper, sulphur, dinocap or thiram.

Grey mould

Caused by a fungus, *Botrytis cinerea,* this is a very common trouble in autumn, when affected plants become covered with a grey, woolly mildew, which soon spreads to other plants. It is so obvious as to be unmistakable. Once it has been seen the affected plant should be disposed of immediately, as there is no real cure. Prevention, however, is fairly easy in that a free circulation of warm air will usually keep the disease at bay, particularly if the plants are given enough water to keep them turgid. Cold clammy conditions, on the other hand, will soon bring on an attack of grey mould.

Sooty mould

This is perhaps the most common fungus disease in the greenhouse and it is all the more confusing to the beginner as it does not look like a fungus. Its effect is to turn the leaf black—hence its name—but fortunately it rarely occurs except on plants that have been disfigured already by the honey dew deposited by sap-sucking insects, for it is on this that the fungus develops. The remedy, obviously, is to keep the plants free from these sap-sucking insects, when sooty mould will cause no further trouble.

Oedema

Known also as dropsy, which generally appears only on zonal pelargoniums (geraniums) and the trailing ivy-leaved kinds. It is not a disease but a physiological disorder caused by the plant's taking up more water than it can give off through its leaves, with the result that the cells of the leaf swell and burst, giving the leaf a corky pimpled appearance. It can be easily prevented by watering the plant less and keeping it in a drier atmosphere.

Chlorosis

There are several possible causes of the yellowing of leaves, generally termed chlorosis. Lack of food, over- and under-watering (particularly the latter), faulty root action and unsuitable soil can all cause it, but it should not be difficult to identify the source of the trouble and to put things right. With calcifuge (lime-hating) plants such as azaleas, heathers and rhododendrons, the usual cause of chlorosis is an alkaline (limy) soil and for these an acid, lime-free soil should always be used.

Scorch

Brown shrivelled areas on the leaves, often including the edges, are usually a sign of scorch, which may be caused in various ways. In spring the young leaves of grape-vines are often affected in this way if the temperature is allowed to rise very high on sunny mornings and the ventilators are then opened too far, causing a sudden drop in temperature. Extremely hot and arid conditions may also produce scorch but a more common cause lies in wetting the foliage with the watering-can or syringe on a sunny day when the ventilators are closed. Wetting the foliage does no harm if the ventilators are open. The sun shining on plants when they have just been sprayed with an insecticide may also cause trouble in this way, while another possible cause is that of fumes escaping from the heating apparatus. In all such cases there is no remedy for the damage but it may be prevented by attending to the points I have mentioned.

INSECTICIDES AND THEIR USE

All the insecticides mentioned in this chapter are readily available in one proprietary form or another, such as dusts, sprays, aerosols and fumigants. With all of them it is important to read the instructions carefully, since information is usually given about possible damage to certain plants.

For small infestations aerosols provide a quick and simple method of control, and the insecticidal dusts too can be easily and conveniently applied, preferably through a powder blower.

With both aerosols and liquid sprays, however, spraying in sunny weather should be avoided if possible. On warm days in summer the job is best done in the evening, as the liquid not only freshens up the plants then but stays on them all night before evaporating in the morning. In winter the reverse applies: the spraying should be carried out in the morning to avoid the plants having to stand wet and cold all night. For the best results a second and third spraying should follow at intervals of a few days so that any insects which hatch out after the first dose are caught by the later applications.

Fumigation

This is undoubtedly the most effective method of controlling insect pests, provided the greenhouse is reasonably free from cracks and other openings through which the fumes can escape. Smoke cones or canisters, which merely have to be lit and left, are available for dealing with most of the different pests and are in sizes to suit any greenhouse; as well as killing the insects, these fumigants coat the plants with a protective film of the insecticide. One drawback of some of the modern fumigants, however, is that they may taint edible crops; and these therefore cannot be gathered for probably a week or so after the fumigant has been used. In this case nicotine or nicotine compounds may be used instead, in one of the special lamps that are available for the purpose or in the form of shreds which can be set alight. Edible crops may then be safely gathered after two days. Nicotine also has the advantage of being harmless to all plants.

As has been pointed out earlier, the greenhouse should be made as leakproof as possible before fumigation is carried out. The job is best done in still weather and preferably at night, provided a temperature of about 65° F (18° C) can be maintained: most fumigants are most effective in fairly warm conditions. The greenhouse can then be locked and left until morning with a notice on the door saying that fumigation is being carried out. As with sprays and aerosols, at least one more fumigation should be done a few days afterwards to deal with any insects that have hatched out from eggs unaffected previously.

Chapter 8

Tomatoes, Cucumbers and Melons

Tomatoes are probably the most widely grown greenhouse crop of all. Although not by any means a difficult crop to grow, it does nevertheless require considerable care if it is to yield the maximum weight of good-quality, well-flavoured fruit, and this means that attention to detail is necessary right from the start.

The tomato is grown invariably from seed, but whether you raise your own young plants or buy them in will depend to some extent on the facilities available. A temperature of about 65° F (18° C) is needed for the germination of the seed, which for an early crop must be sown in January or early February; and something approaching this temperature must also be maintained until the plants are established in their final quarters. Unless there are other plants needing relatively high temperatures, it usually pays therefore to buy in the plants for an early crop. For a later crop it should be safe to sow the seed in early March, when the necessary temperatures can be maintained more easily and when other plants will probably need them anyway. In either case it is essential to remember that it is no use trying to raise the plants in cooler conditions than those mentioned—tomato roots do not grow at less than 57° F (14° C)—for even if the plants survive they will probably receive a check from which they will never really recover.

If plants are to be bought in, it is most important to get good sturdy ones. Look for thick short-jointed stems and leaves of a good green colour. There should also be plenty of healthy white roots and preferably a pair of seed-leaves at the base of the stem. These incidentally are quite different in appearance from the ordinary tomato leaves and their presence is a sure indication that the plant has received no check. The plants to avoid are those with thin 'drawn' stems, yellowish-green leaves widely spaced apart and no seed-leaves.

Seed sowing

If the plants are to be grown from seed, this should be spaced out 1in apart and ¼in deep in either the J.I.P1 or a soilless compost, thus reducing the risk of damping off. Kept warm and moist the seeds should germinate in about ten days, when the seedlings must be given maximum light to keep them short and stocky. Then as soon as the seed-leaves are touching, the plants must be pricked out into either the J.I.P2 or a soilless compost. Better still, they can be potted up at once into 3in pots. Whichever method is used, they must be handled by the leaves (not by the stem, which is very vulnerable) and buried almost up to the seed-leaves.

Seedlings pricked out into seed trays must be potted up into 3in pots as soon as their leaves are touching and from then on the potted plants must be kept warm and moist in full light.

Preparing for planting

Tomatoes can be grown in the actual soil of the greenhouse; in pots or boxes; or in bottomless pots (ring culture). There is also a method of growing them on straw bales, but as this is of more interest to the commercial grower than to the amateur, I shall not deal with it here.

If the greenhouse has staging with glazed sides above it then pots, boxes or ring culture will have to be used, as it is most important to keep the plants as near as possible to the light. Where the glass goes almost to the ground, planting in the actual soil can be carried out, but although this vastly reduces the amount of watering needed it also brings increased risk of soil-borne disease. Ring culture at ground level is therefore usually a better proposition.

Pots or boxes should be amply large enough, with perfect drainage: wooden orange boxes are ideal. Pots should be not less than 9in. Whichever type of container is used, it should be filled evenly and firmly with the J.I.P3 to about 2in from the top.

For ring culture special bottomless pots of tough paper are

commonly used. These are filled in the same way as the pots or boxes but they are first stood on a base of aggregate, which may be weathered ashes, clean clinker or shingle. If this base is to be at ground level, the soil should be removed to a depth of 4in or 5in and replaced with the aggregate, while if ring culture is to be carried out on the staging a similar bed of aggregate should be made up on it.

For plants to be grown actually in the greenhouse soil, this will have to be prepared thoroughly by making sure that drainage is perfect and digging the soil over to a depth of 18in, keeping the top soil at the top and working in plenty of old manure or garden compost. The digging and manuring are best done in the autumn, or at any rate as soon as the ground becomes available. Then during the winter these operations should be followed by a thorough soaking that wets the soil right through to the bottom. Finally, in spring, a tomato base fertiliser (obtainable at garden shops) is dug in a short time before planting takes place. Dry bonfire ash, including charcoal, is a good material to mix with the soil, as it helps to produce high-quality fruit of good flavour.

Planting

This should not take place until the soil or compost is quite warm and reasonably moist but not wet. The young plants too should be moist at the roots before they are put in. The actual job is best done with a trowel, setting the plants 18in apart each way with the top of the soil-ball from the pot just level with the surface. The adjacent soil is then firmed gently round the soil-ball until there are no spaces and the soil is of roughly the same firmness as the soil-ball.

For the first watering only enough should be given to settle the soil around the roots. From then on the soil-ball should be kept moist, with the amount of water being increased gradually as the roots spread. The two things to be guarded against at this stage are dryness of the soil-ball, which may result in a poor root system that will not be able to support the mature plant; and over-watering, which will produce large soft growth with possible loss of the first truss of fruit.

Subsequent treatment

Tomato plants are generally grown with a single stem and this means that all the sideshoots produced in the axils of the leaves (where the leaf joins the stem) must be pinched out as soon as they are large enough to handle.

Fig 3 The sideshoots growing in the axils of tomato leaves must be pinched out while quite small

The plants will also need supporting and for those grown in the ground a simple method is to use the soft string marketed as four-ply fillis. This should be looped round the base of the stem and carried up to wires running along the greenhouse roof and supported in screw-eyes. The plants are then merely wound round the string in a clockwise direction.

For plants on the staging, wires may be run horizontally along the slope of the roof, so that canes can be tied to them with the plants in turn being tied to these. Constant tying is needed as otherwise the plants will grow towards the glass, resulting in crooked stems which, being brittle, cannot be trained back on to the canes.

A common practice is to stop the plants by pinching out the tip of the stem above the fourth or fifth truss; but if they were put in early and are doing well, there is no reason why they should not be left to reach the top of the supports provided watering and feeding are not neglected.

From the time the first flowers open, overhead damping should be carried out to keep the pollen of the flowers in good condition and also to dislodge it. Otherwise it may become too dry to function properly or may not come into contact with the other flowers, with the result that pollination will not take place and the fruit will not set. In commercial nurseries this overhead damping is done by spraying the plants forcibly with a hose, but in the amateur's greenhouse a watering-can or syringe can be used instead to wet the plants thoroughly once or twice a day.

It is quite safe to do this even in full sun as long as ample ventilation is provided at the same time. If it is not done, a trouble known as 'dry set' may occur, with the fruits either not forming or growing no larger than cherries—a common trouble on the bottom truss.

Ventilation should in any case be given freely, as long as this can be done without exposing the plants to draughts or cold winds. The tomato thrives on fresh air and towards the end of the summer ventilation is even more important, as without a free circulation of air there will probably soon be an attack of grey mould (see Chapter 7), with the result that many of the fruits will fall.

Watering and feeding

Once the first truss of fruit has set, the plants will need copious supplies of water. For plants grown in the ground this must be given in sufficient quantities at one time to reach down to the lowest roots. Usually the time to apply it is when the soil is showing signs of drying out on the surface.

If the plants are grown in pots or boxes almost daily watering will be needed: make sure each time that the compost is soaked right through. In hot weather watering may even be necessary twice a day.

In ring culture, once the plants are well established, watering is limited to the aggregate. This will need wetting thoroughly every day or even more frequently in hot weather.

Feeding and watering are usually done at the same time and a start should be made as soon as the bottom truss of fruit has set, using a proprietary tomato fertiliser at the maker's recommended rate. This will probably include a fairly high proportion of potash, which is valuable in the early stages of the plant's growth; but by the time the fourth truss has set, a more nitrogenous feed is usually needed as the plants are then putting all their effort into the fruit, with the result that growth slows up considerably. A dressing of dried blood is as good as anything at this stage.

Feeding may also need to be varied according to the weather. More potash, in the form of sulphate of potash, is necessary in dull grey summers, while more nitrogen is advisable in hot dry ones.

In the case of ring culture, the feeding of plants is restricted entirely to the compost in the rings, and owing to the limited root space rather heavier feeding tends to be needed. As a rough guide, 3pt of liquid fertiliser should be applied to each ring once a week.

Tomato cultivars

Some cultivars do better than others on different soils and with different methods of cultivation. However, a good strain of the long-popular Ailsa Craig is still generally regarded as being one of the best for cropping and flavour, while the more modern F_1 hybrids such as Eurocross, Kingley Cross, Ware Cross and Maacross, are being grown increasingly. (F_1 indicates that the seed is the 'first cross' or first generation from parent plants which have been specially selected and grown to give extra strength and vigour to the progeny. The term is used in connection with several other plants from seed, and the seed is always considerably more expensive than the normal sort.)

For those who fancy a novelty, there are several unusual cultivars. Golden Sunrise and Sutton's Golden Queen are yellow-fruited types of excellent flavour. Then there are Tangella with tangerine-coloured fruits and the white-fruited White American Beauty. Finally, there is an Italian cultivar, Ponderosa Scarlatto, with huge fruits that come late in the season. In fact it is claimed that this can be gathered after most others have finished.

Tomato troubles

Tomatoes are beset by all sorts of disorders and diseases but fortunately these are not as prevalent as they might be. The fact that tomatoes are grown so extensively is an indication that they are relatively trouble-free, but even under the best cultivation problems can sometimes arise and a few of the more common disorders are therefore described below.

Blossom end rot

Occasionally, some of the first fruits have black areas inside them. These appear eventually on the surface, usually at the outer end of the fruit where they are seen as sunken, dark brown patches. This is not a disease but a physiological disorder caused by insufficient water reaching the fruit, most likely as the result of a too dry soil: it seems that when there is a shortage of water, the foliage takes precedence over the fruit.

Blossom drop

This trouble is related to the dry set mentioned above in connection with the overhead damping of the plants. Instead of the fruit failing merely to set properly, the whole flower including its stalk breaks off at the 'knuckle' near its base. The cause is usually a too dry soil, although the trouble also occurs frequently at the top of heavy-cropping plants as a result of poor root action.

Fruits falling off

Tends to occur at the very end of the season, when some of the fruits may go soft and discoloured where they are attached to the calyx. The trouble is always caused by grey mould (see Chapter 7).

Leaf mould

Where leaf mould occurs consistently, it is best to try one of the resistant cultivars that are now available. Caused by the fungus *Cladosporium fulvum*, the disease starts as a grey mould on the underside of the leaves, which gradually turn purple while the upper sides become orange-yellow. Eventually the leaves become covered with the fungus and dry up and die. The trouble affects both leaves and fruit and is usually worse where both temperature and humidity are high. A partial control therefore can be obtained by providing a cooler and drier atmosphere, although at the first sign of an attack the plants should be sprayed with one of the copper fungicides that are readily available.

Blotchy ripening

Not a disease but a disorder in which, as the name implies, the fruits ripen unevenly, generally with yellow or orange patches. It is caused usually by a lack of potash or, if the supply of this is adequate, by a poor root system that cannot make full use of it.

Wilting

This may be due to no more than a poor root system, particularly if the plants merely wilt in the sun and pick up again in the evening. There is, however, a far more serious form of wilting known as verticillium wilt, which necessitates the burning of the affected plants and the sterilisation or renewal of the soil in which they have been growing. Unlike the wilts produced by poor root action, verticillium wilt starts at the bottom of the plant and works its way up, until eventually the foliage becomes a blotchy yellow and the plant collapses and dies. The best test for the presence of this disease is to take up the suspected plant and to split it in two along the length of its stem, when if the wilt is present there will be a discoloration of the inner tissues along almost the entire length. Although this may mean wasting a plant if the disease is not present, it is better than running the risk of allowing the wilt to spread through the whole crop.

CUCUMBERS

It is seldom satisfactory to grow tomatoes and cucumbers together in the same greenhouse as the latter need a far more humid atmosphere, although it may be possible to get over this by screening off the cucumbers with polythene sheeting.

Plants from seed

Cucumbers need a good deal of warmth in their early stages and it is not advisable to make a start before about mid-March, unless the greenhouse can be kept to a minimum of 60° F (16° C). For the germination of the seed an even higher temperature, 70–75° F (21–24° C), is needed. This nearly always necessitates the use of a propagating frame, preferably with bottom heat.

The seeds are sown singly in 3½in pots and set horizontally about ½in deep in the J.I.P1. If the compost is quite moist and warm to start with, no further watering will be needed; and if the temperature mentioned above is maintained, germination should take place within three or four days. Careful watering and ample humid warmth should then keep the plants growing well until they are about 6in high and ready for planting out.

Planting

Cucumbers may be grown in 10in pots of J.I.P3 but a better method is to prepare a special bed for them. This may be made on asbestos or corrugated iron sheeting laid on the staging; or if the greenhouse is glazed almost to the ground, it may be prepared on the actual soil as long as this is well drained.

To make the bed, put down a 6in layer of 2 parts rotted turf and 1 part old strawy manure, with a good sprinkling of bonemeal and a lighter dressing of lime worked in. Then for each of the plants, which should be 3ft apart, build up a mound of a similar mixture on the bed using, say, about a bucketful for each mound.

The next step is to get the bed and the mounds quite warm before planting, but with both artificial and sun heat it should not take long to bring them to the required temperature of 65° F (18° C). The young plants, which should have been given a good watering first, are then put in so that the top of the soil-ball is about ½in above the top of the mound.

From then onwards the soil must be kept constantly moist with water at the same temperature as that of the greenhouse and later on, when roots appear on the surface, a top dressing of the same soil mixture should be applied about 1in thick, this being repeated each time the roots appear again. Frequent damping down, without wetting the cucumber leaves too much, is also necessary in order to maintain the humid atmosphere and to keep down red spider mite.

Training the plants

Cucumbers are trained on horizontal wires, running along the slope of the greenhouse roof and about 12in from it, with about 9in between the wires. Alternatively, one of the modern forms of rigid wire mesh might be used, similarly held away from the glass.

The plants should be allowed to grow until they reach the top of the supports, when they should be stopped. In the meantime sideshoots will have grown from the axils of the leaves, and these should be stopped as soon as they have made two leaves beyond the tiny cucumber which should form on each of them. The secondary sideshoots that will grow on these should then be stopped likewise.

All male flowers (those without an embryo cucumber behind them) must be removed, as they will otherwise pollinate the female ones and result in the production of seed. If this happens, the seed-bearing cucumber will develop a swollen end, something like an Indian club, and be far less palatable.

Cucumbers in frames

Plants for frame culture may be raised in the same way as those for the greenhouse but a start should not be made until the second half of April, to provide plants for putting in the frame in early June. For the best results a frame 6ft by 4ft is needed, and one plant should be set at the centre of this on a bed of rich soil. As soon as the plant has made six leaves, it should be stopped so that

Page 69 (*above*) Pots must be well crocked for drainage, using pieces of pot placed concave side downwards over the drainage hole; (*below*) young gloxinias being carefully lifted out of the seed tray, ready for potting up

Page 70 (*above*) Hydrangea cuttings being inserted round the edge of a pot; (*below*) indoor ivies can be layered easily by pegging a stem down into a small pot

four sideshoots can be trained out, one to each corner of the frame. Subsequent sideshoots from these should then be stopped as soon as they reach the sides of the frame. Careful attention to watering and ventilation will be needed to keep the plants growing in a humid but buoyant atmosphere: in hot weather the frame will have to be shaded.

Cultivars

Comparatively few cultivars of cucumber are grown. Two of the most popular for the heated greenhouse are Improved Telegraph and Butcher's Disease Resister. For the cold greenhouse or frame, Conqueror is more suitable.

MELONS

The early stages of melon cultivation, from sowing to planting, are very much the same as for cucumbers, and much of the subsequent treatment of watering, damping down, top dressing and ventilation is also similar.

It is usual to train the plants on horizontal and vertical wires forming a trellis of 9in squares up the slope of the greenhouse roof. The plants are allowed to grow up this until they have reached about 4ft, when they should be stopped. While the plants are growing to this height, sideshoots will be produced. These are trained along the horizontal wires, stopping them at about 12in long.

A plant grown in this way will carry four fruits, each on one of the sideshoots, the remainder of the sideshoots being removed. As with cucumbers, there are male and female flowers, and it is on the latter that the embryo melon develops behind the bloom. Unlike those of cucumbers, however, these female flowers must be pollinated by the male ones. The best way to do this is by picking a male bloom, turning back its petals to expose the stamens, and inserting it into the female bloom. This should be done preferably about noon on a sunny day, when all the flowers on any one plant should be dealt with at the same time. The fruits that have been pollinated successfully will then swell, and as soon as four good specimens can be selected, the remainder should be removed. Each sideshoot carrying a swelling fruit should be stopped one leaf beyond the fruit.

The developing fruits are best supported in the special melon nets that are available and once they have reached their maximum size (indicated by the cracking of the stalk or skin), all watering must cease.

The fruits are ripe when they are slightly soft at the base, with a noticeable aroma. They should then be gathered by severing the shoot on which they are growing 2in above and 2in below the melon stalk.

Cultivation in frames

Two melon plants can be grown in a 6ft by 4ft frame. In this the young plants should be set out at the end of May on a slightly mounded bed of rich soil. After the plants have been stopped at the fourth leaf, the resultant sideshoots can be allowed to grow until they reach the sides of the frame, when they must be stopped. The treatment then is the same as for greenhouse melons, but only two or three fruits should be allowed to grow on each plant.

Cultivars

For a heated greenhouse, Hero of Lockinge and King George are suitable; while for cold or only slightly heated structures, Tiger and Dutch Net are better. These last two may also be grown in frames.

Chapter 9

Grapes, Peaches and Nectarines

Grapes may be grown in heated or cold greenhouses or conservatories provided these get plenty of sun. The great drawback with this fruit, however, is that the vines create a good deal of shade in summer, while in autumn the fallen leaves can lead to trouble among the plants growing beneath. Grapes are therefore best cultivated on their own in a house or 'vinery', but as this is impractical for most people, a compromise might be reached by growing beneath the vines only those summer plants which do not mind shade: fuchsias, gloxinias, ferns and *Begonia rex*. All these can be put out of the way by the time the vine leaves fall. Tomatoes seldom do well beneath vines owing to the poor light and fairly humid atmosphere; but most spring-flowering pot plants, including bulbs, are usually successful as they are largely finished by the time the vines start into growth.

Vine borders

Although grapes can be grown in pots or tubs, the more popular method is to use a prepared bed of soil either inside or outside the greenhouse. For early grapes that are produced on vines started into growth during the first part of the year, an inside border is usually preferable but for a later crop a common practice is to plant the vine close against the outside of the greenhouse wall and to bring the main stem in through a hole low down in the wall. By doing this much of the labour of watering is avoided.

Whichever method is used, the drainage must be perfect. If there is any doubt about this, the soil should first be dug out to a depth of about 2½ft, with a good layer of rubble then placed in the bottom and connected to land drains that will lead away any excess water. In these circumstances it is also advisable to line the hole with bricks, slates or asbestos sheeting before refilling it with soil, as this will prevent the roots from spreading into unsuitable soil. A bed about 12sq ft in area will take one vine, which is all that a small greenhouse will accommodate.

Any good fertile soil will grow grapes well, but even the best will be improved by the addition of plenty of garden compost or well-rotted manure, plus 1lb of bonemeal and 1oz of sulphate of potash per square yard. A dressing of lime, at about ½lb per square yard, is also generally advisable. If the soil is very poor, better results will be obtained if it is replaced with rotted turves to which the above materials have been added.

Planting

This may be done at any time during autumn or early winter, when pot-grown 'canes' may be purchased. Make sure that the soil in both the pot and the border is moist, then take the plant carefully out of the pot so that no roots are damaged: the best way to do this is to break the pot. The roots should then be gently disentangled so that they can be spread out in the planting hole and covered very firmly with about 2in of soil, although the actual base of the cane should not be buried more than an inch or so. As mentioned above, vines planted outside may be set quite close to the wall, but when they are planted in inside borders they should be 12–18in away from it.

Pruning

Immediately after planting, the cane should be cut back to ripe wood (which is easily recognised by its hard, yellowish-brown appearance), although this may have been done already at the nursery. The cut should be made just above a joint, since if a portion of stem is left above the joint it will only die back. The remaining portion of cane should then still be long enough to be brought into the greenhouse through a hole in the wall if the vine has been planted outside, or in the case of one planted inside it should be tied to a 6ft can, firmly inserted in the ground.

This, and all subsequent winter prunings, must be carried out soon after the leaves have fallen and when the vine is quite dormant, since if it is done when the buds have already started into growth, it is probable that the vine will 'bleed' by exuding sap from the cut surfaces. If this happens it is difficult to stop it, but covering the wounds with painters' knotting or sealing wax is reasonably effective.

During the spring following planting, new shoots will appear and the top one should be allowed to grow upwards unchecked: this is the one that will form the main stem or 'rod' on which the grapes will be produced for many years. All the other shoots or laterals should, however, be stopped when they have made six leaves and the sublaterals from these should in turn be stopped at the first leaf.

All this growth needs to be pruned quite drastically in the winter, when the extension growth made by the main stem in summer should be shortened back to about one-third of its lengths. All, the side shoots should be pruned hard back to one or two buds which usually means little more than ½in from the main stem. It is these tiny stubs that will eventually develop into the 'spurs' on which the fruit will be produced annually.

In cool houses the vine will probably not start into growth until March. If a little heat is available it may be started into growth in February, but no attempt should be made to force it along in high temperatures.

The first sign of starting will be the breaking of the buds on the stubs of the previous year's sideshoots: one or two shoots normally develop from each stub. The usual practice is to reduce these to one shoot per stub as soon as the strongest on each can be selected, but for the beginner a safer way is to allow all the shoots to grow until the fruit clusters can be seen, when the shoot bearing the best cluster should be retained and all others on the stubs removed completely. The fruit-bearing shoot should then be stopped two leaves beyond the bunch, while all the other laterals and sublaterals which will be produced on the new extension growth at the top of the vine should be pruned as described for the first year's growth.

The spurs develop as the result of the annual winter pruning of the fruited shoots, which are always taken back to the buds at their base; thus what started as a mere stub becomes in time a compact, much branched growth. When this becomes inconveniently large it may be reduced in size, but before then it may also be advisable to cut out some of the spurs altogether, as one spur to every 12in of rod is quite sufficient.

Setting and thinning the fruit

Damping down to create a fairly humid atmosphere should be carried out during most of the summer, except when the vine is in flower. Rather drier conditions are needed then to encourage the fruit to set. Setting will also be assisted by tapping the vine rods to dislodge the pollen from the flowers. With this treatment most of the commonly grown vines will set freely and the next task is to thin the fruit.

This is a tedious job involving the removal of many of the berries to leave room for the others to develop properly. It should be started as soon as the berries are nearly ¼in across, when those near the centre and towards the base of the branchlets forming the cluster should be carefully removed with a pair of pointed scissors (special grape scissors are available for this). To avoid handling the fruit, a thin forked stick should be used to support the fruit stalks as the berries are cut out.

It is impossible to give any guide as to the actual spacing of the berries after thinning has been completed, but they should be wide enough apart to allow a free circulation of air and to give each berry ample room in which to develop to a good size.

General treatment

When the vine is starting into growth, ventilation should be just sufficient to maintain a free circulation of air, except on sunny days when the temperature should be kept below 65° F (18° C) by admitting more air. Then during the summer, ventilation should be given freely until the grapes are ripening, when the aim should be to retain the sun's warmth and at the same time keep the air on the move.

The watering of indoor vines will also need attention, and

whenever the soil shows signs of drying in the spring and summer it should be soaked thoroughly. Mulching with leaf-mould, garden compost or rotted manure will reduce the need for watering and also supply a certain amount of food, but for the best results a proprietary vine fertiliser should be applied in April.

In winter the vine must be kept as cool as possible, and during its winter rest in the first year the necessary supports for it can be provided. These usually consist of lengths of No 8 gauge wire suspended on vine-eyes screwed to the greenhouse roof (these are obtainable at horticultural suppliers). The wires should be spaced about 12in apart to cope with the main rod and the sideshoots.

Vine cultivars

For the cold or slightly heated greenhouse, Black Hamburgh is by far the most popular and easily grown cultivar, its only drawback being that it is a fairly late cropper. For an earlier crop in heated houses, Foster's Seedling is a good choice.

Vine troubles

Powdery mildew and grey mould sometimes attack vines—the latter particularly if the berries have not been thinned enough—but these have already been dealt with in Chapter 7. The berries may also be troubled by shanking, which causes the stems of the affected berries to become black and withered with the results that the fruit is shrivelled and sour. The cause is usually poor root action leading to starvation and unhealthy growth, with poor cultivation in general being a contributory factor. Prevention lies largely in ensuring that there is ample drainage and food and that the vine is not overcropped or badly pruned.

Sunscald, which causes the berries to become brown and sunken on the side facing the sun, is generally due to the berries being moist and without adequate ventilation when the sun is shining on them. It should not occur in well-ventilated houses.

Propagation

The grape-vine is easily propagated in winter by using 'eyes', each consisting of a piece of mature woody stem about 2in long, with a dormant bud at its centre. The eye is first prepared by removing a sliver of wood about ⅛in thick from the side away from the bud; it is then inserted in a 3in pot of sandy soil so that the bud is just above the surface. Kept warm and just moist in a propagating frame the bud will soon start into growth, when the plant should be removed to cooler conditions and grown on until the pot is well filled with roots. It can then be planted where it is to remain.

Young vines raised in this way should be kept growing freely through the summer until they are pruned back to the ripe wood in the early winter; then the following year it should be possible to take one or two bunches of fruit from each vine.

PEACHES AND NECTARINES

Both peaches and nectarines are commonly grown in cold or heated greenhouses, largely because their flowers, which are produced early in the year, are easily damaged by frost. Under glass, the most practical position for the trees is against the back wall of a lean-to greenhouse, where it is usual to grow them on wires as fan-trained, but they may also be grown as bush trees in large pots.

Soil and planting

Planting is done between October and the end of February, but the soil should be prepared first by making sure that it is perfectly drained and not short of lime. Heavy soils can be improved by the addition of weathered ashes or coarse sand, while very light ones will need plenty of compost, leaf-mould or damp peat to retain moisture. A liberal dressing of bonemeal should also be worked in.

The ground must be made quite firm before the plants are put in, when it is important to ensure that the point of budding or grafting (indicated by a swelling on the trunk) is well above the surface. When the tree is grown against a wall, the base of the trunk should be about 6in away from it, with the top growth sloping slightly inwards towards the wall. Wires to support fan-trained trees will be needed but the trees should be tied only temporarily to these until the ground has settled, say in a few months' time. After planting, give the plants a thorough soaking, then make sure that the roots never become dry. If more than one fan-trained tree is planted, at least 15ft should be allowed from one to the other.

Pruning

Both the peach and the nectarine need more continuous pruning than almost any other fruit and the operation can be a fairly confusing one to the beginner. It will, however, be understood more easily if it is appreciated that the best fruit is produced on the previous year's growths; that is, a shoot that grows one year will produce fruit in the next.

The initial training of fan-trained trees is a rather complicated business best left to the nurseryman and for this reason it is advisable to start off with three-year-old trees that have already been trained to shape.

During the summer following planting, new shoots will be produced both along the main branches and at their ends, those in the latter position being the extension growths which will lengthen the existing branches. These can be left to grow during the summer, tying them in to the wires in the case of trees grown against walls, but any sideshoots from them should be pinched back to one leaf. Then in winter the extension shoots themselves should be cut back to three or four buds.

It is, however, the shoots produced along the length of the main branches that are the important ones, since these will include the fruiting ones for the following year. The first job is to select these. The firmest and strongest shoots are the ones to keep and these should be spaced about 9in apart, with all others being either pinched hard back or removed entirely unless they are needed to form new branches for filling in gaps. These selected shoots are then tied in and eventually stopped when they are 12–15in long. They are left at this length and by the winter they should have developed into fruiting wood bearing both fruit and growth buds, the former being quite rounded in shape and the latter long and narrow.

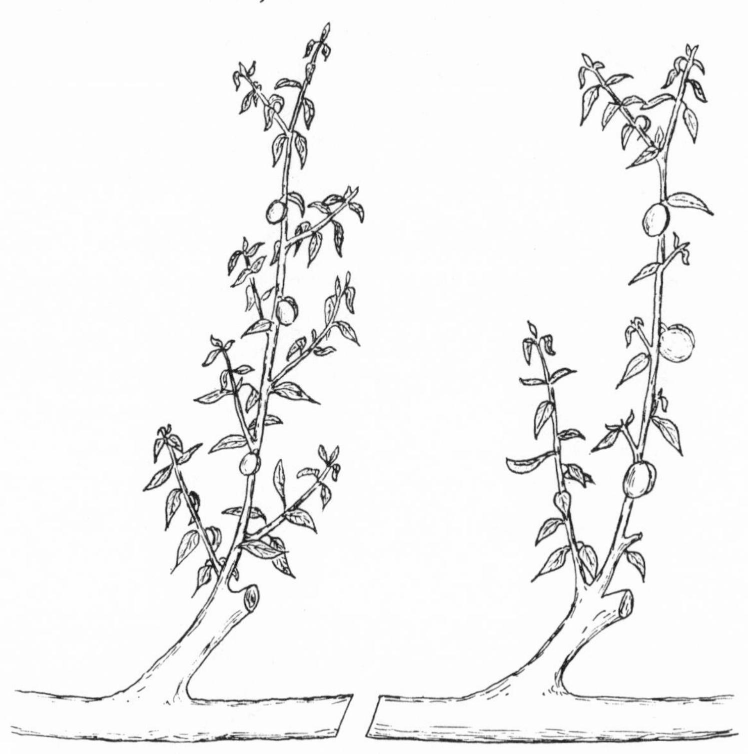

Fig 4 Disbudding peaches. On the left is shown a fruiting shoot before disbudding; (right) a disbudded shoot, with all sideshoots pinched out except one at the base to provide an eventual replacement shoot and one at the top to draw up the sap through the fruiting stem

In spring two shoots, one at the tip and the other at the base, are allowed to grow from each of these fruit-bearing ones, with all others being removed gradually: this is the process known as disbudding. The idea of leaving a shoot at the tip is to encourage a flow of sap that will help to feed the fruits developing lower down, but the shoot at the base is actually the more important of the two as this is the replacement shoot which will fruit the following year. It is a good policy to leave two shoots at the base to start with, in case one becomes damaged; then when the two are a few inches long, the stronger one should be retained.

The shoot at the tip should be pinched back to about halfway when it has made eight leaves.

After the fruit has been gathered, the fruited shoot is cut right out, leaving the replacement shoot to develop and fruit in the same way the following year. Then in turn this is cut out and replaced with another one from its base.

Setting and thinning the fruit

Peaches, including nectarines, are self-fertile and if the greenhouse is well ventilated, bees and other insects may pollinate the flowers sufficiently; but it is advisable to hand pollinate as well, using a rabbit's tail to transfer the pollen from one flower to another.

Even without this, fan-trained trees often set more fruits than they will be able to mature properly and a certain amount of thinning is therefore necessary. This should be done as soon as the fruits are as large as hazel-nuts, when the number on each fruit-bearing shoot should be reduced to two. A further natural thinning is sure to take place as the fruits form their stones. When they start to swell again after stoning, they should be reduced still further to allow about 1sq ft to each in the case of peaches and rather less for nectarines.

Finally, when the fruit is ripe, gathering must be done with care as even finger pressure will bruise the fruit. To test for ripeness, each fruit should be cradled in the palm of the hand and lifted gently. If it is ready it will come away in your hand.

Peach leaf curl

This is a fairly common trouble, although less so under glass than outside. On affected trees the leaves are seen swollen, distorted and red before they become covered eventually with a white 'bloom' prior to falling prematurely. Spraying with Bordeaux or Burgundy mixture (obtainable at good garden shops), when the buds are just starting to swell in spring, will usually give effective control.

Chapter 10

Miscellaneous Fruits and Vegetables

Although it is hardly an economic proposition to grow the more exotic fruits under glass there are several that are worth trying, particularly as most of these make very decorative plants apart from their fruit. In general their cultivation is even easier than that of the vine or the peach, as little in the way of elaborate pruning is needed. The only real difficulty lies in ripening the fruits in a climate that is much cooler than their natural one.

THE CITRUS FRUITS

Two or three centuries ago it was fashionable for every stately home to have an elaborately constructed 'orangery' in which the various citrus fruits were grown. But gradually the orange and its relatives faded into oblivion as cultivated plants in Britain, and nowadays it is difficult to find plants of more than a very limited number of different kinds.

Nevertheless, the few that are available make interesting plants to grow. Their glossy evergreen leaves alone make them handsome subjects; and they are further enhanced by their very fragrant white flowers and by the fruits which often appear at the same time as the flowers. For one of the peculiarities of these plants is that in cool conditions the fruits produced in one summer do not ripen until the next, so that it is not unusual to have flowers, ripe fruits and unripe fruits on the tree all at the same time.

It is no use trying to grow the trees from pips. Although these will germinate and grow easily enough into attractive foliage plants, there is no chance of their flowering and fruiting for many years, and even then it will be at a size that would prohibit their cultivation in the average greenhouse or conservatory.

For trees that will flower and fruit at an early age and a convenient size, shoots of named kinds are budded or grafted on to seedling stocks, although cuttings may also be used if a fruiting tree is available to supply them. The usual way of starting is to buy young plants, of which the most easily obtainable are the calamondin *Citrus mitis*, a recent introduction with small fruits, generally grown as a decorative house plant; the Chinese Orange *C. taitensis*, with fruits the size of a tangerine, excellent for marmalade; *C. limonia ponderosa*, which in a large tub will grow to 8ft and produce enormous lemons; *C. limonia meyeri*, another lemon, smaller in every way than the last; *C. paradisi*, the grapefruit, suitable for a large tub; and the Shaddock *C. maxima*, with enormous pear-shaped lemon-yellow fruits more pungently flavoured than the orange.

Cultivation

A suitable compost for all the citrus family is a medium heavy soil to which some compost or old manure, plus a good sprinkling of bonemeal, has been added. As a rule it will be found more convenient to grow the plants in pots or tubs than in the ground but these containers must be well drained and the compost must be made very firm, otherwise in a too loose soil the annual shoots may not ripen to the fruiting stage.

To save space the plants intended for tubs may be grown in large pots for the first two or three years, together with the smaller species, such as *C. mitis* and *C. taitensis*, which will remain permanently in pots. Repotting is not needed very often as the plants do best when their roots are confined, but if they make little progress early in the year, when they should be growing vigorously, this is generally an indication that repotting is required.

During the summer, liberal watering and syringing are needed together with generous feeding, and from early June to the end of August the plants should be stood outside in a sheltered position. When they are brought inside for the winter they should be kept in a temperature of 45–50° F (7–10° C) and given only enough water to prevent the compost from drying out; but when watering is done it must be done thoroughly. An occasional dose of Epsom salts (magnesium sulphate) is also helpful as the citrus plants soon suffer from magnesium deficiency.

Very little pruning is needed beyond the removal or cutting back of any badly placed or straggling shoots, but hand pollination of the flowers using a rabbit's tail is advisable to ensure a good set of fruit. A light shading should also be provided when the plants are under glass in early summer, but at other times of the year they should be given as light a position as possible.

Mention might also be made here of the pomegranate *Punica granatum*, which superficially resembles the citrus fruits enough to mislead the amateur into thinking that it is one of them. It is, however, a quite different plant belonging to another family. Although it is hardy enough to grow and flower outside in Britain, it rarely ripens its fruits; but there is a possibility that it might do so under glass in a good summer in mild areas. It is well worth growing for its scarlet flowers, each about 1½in across. These are produced from midsummer onwards on a tree which, unlike the citrus fruits, loses its leaves in winter. In severe winters plants grown outside may be cut to the ground, but even so they will usually spring up again from the base. The plants are raised easily from pips which are sown as soon as they are ripe in small pots. If they are not wanted after this for outside, they can be grown on to make decorative pot plants.

THE GRANADILLAS

Valuable for both their flowers and their exotic fruits, the granadillas or passion-fruits are not difficult to grow provided adequate space and heat—at least 60–65° F (16–18° C) in winter—are available.

The two sorts usually grown for fruit are *Passiflora quadrangularis* and *P. edulis*, both of which may be planted either in a very firm fertile soil in the greenhouse border or in boxes or tubs of the J.I.P2. Perfect drainage is essential.

As both plants make very vigorous climbing growth, some means of training them over the greenhouse walls and roof will be necessary. This is a job that will need constant attention, otherwise the stems will soon be in a hopeless tangle. During the summer, and particularly during the fruiting period, very high temperatures will be appreciated by the plants provided watering and damping down are also attended to, for unless they have plenty of root and atmospheric moisture the fruits tend to drop.

On both species the flowers are produced on shoots of the current year and it is advisable to hand pollinate them with a small camel-hair brush. On *P. edulis* there should then be a good crop of 2in long, oval fruits, changing from yellow to a wrinkled purple as they ripen, while on *P. quadrangularis* the fruits are more like large citrus fruits, yellowish-green, oval and up to 8in long.

The succulent pulp of both fruits makes very good eating, but on *P. quadrangularis* particularly there is an additional bonus in the very showy flowers, each 4in or more across, with white petals and green and pink sepals. Above the petals stands the impressive corona of curled filaments, banded blue and reddish-purple. The flowers of *P. edulis*, although smaller and less showy, are still quite attractive with their purple-banded filaments standing above predominantly white and green petals and sepals.

AUBERGINES

Also known as egg-plants, these are not grown as commonly as they might be, for they are not difficult to cultivate and they do not take up much space. The fruits, large and variously shaped, are generally purple but there are also white forms and both types can be eaten either raw or cooked.

Sow the seeds in a temperature of at least 65° F (18° C) early in the year and pot the seedlings on, first into 3in pots of J.I.P1 and then into 7in ones of the J.I.P2. When the plants are about 6in high, pinch out the tips of the main stems to encourage branching and give plenty of water and warmth throughout the summer. Staking and tying will also be needed to prevent the bushy plants from flopping over, and no more than six fruits should be allowed to grow on each plant. The fruits should be gathered when they are a good size and slightly soft.

THE LOQUAT

The orange-yellow, pear-shaped fruits of the loquat *Eriobotrya iaponica* seldom mature out-of-doors, except perhaps in the very mildest parts of England, but they can be ripened successfully under glass, even in a cold greenhouse, which should be preferably of the lean-to type so that the tree can be planted against the back wall. Apart from its fruit it makes a most attractive tree, with clusters of yellowish-white, fragrant flowers set against large, glossy, evergreen leaves.

Once planted in any reasonably good soil, it presents little trouble if it is watered liberally in summer and rather less in winter, while all the pruning that is needed is the trimming back of any straggling growths in spring. Frequent syringing will also help to keep it healthy.

FORCED STRAWBERRIES

Whether strawberries forced under glass are worth all the trouble involved is a matter of opinion, but they make an interesting crop. To make the growing worth while, at least a dozen plants will be needed. At an average of nine fruits to each, these should be sufficient to provide more than a taste at any time from early April onwards, according to the heat available.

For forcing, the old Royal Sovereign is still the best strawberry to grow, but owing to its susceptibility to disease it is essential to obtain clean stock from a specialist grower. Pot-grown plants are usually obtainable in late July or early August and as soon as they are received they should be potted up immediately into well-drained 6in pots of J.I.P3, care being taken to make the compost quite firm and not to bury the plants too deeply: the basal crown should be just 'sitting' on the surface.

After potting, plunge the pots to the rim in a bed of peat, sand or weathered ashes in full sun in the open and keep them quite moist. As soon as growth ceases at the end of the summer, remove them to an airy cold frame, where they must be kept as cool as possible and quite dry except for an overhead syringing on sunny days. As soon as this produces signs of new growth, some

Page 87 (*above*) Most cuttings, such as the zonal pelargonium shown here, must be trimmed off just beneath the joint (*left*) zonal pelargoniums (geraniums) grown indoors should be cut hard back in spring

Page 88 (*above*) Pot saucers provide an easy way of ensuring that the soil is wetted right through; (*below*) young cyclamen buried to the pot rim in a plunge bed for the summer

protection in the form of mats or sacking will be needed on cold nights, but this new growth must still be maintained by keeping the compost just moist, more water being given as the plants develop. As soon as they are growing strongly they should be transferred to the greenhouse, where they can be accommodated conveniently on shelves near the glass.

At this stage a temperature of 45° F (7° C) is adequate and in this the plants should be kept quite moist at the root and syringed frequently overhead. As soon as the flowers open, syringing should cease and rather more warmth should be given.

To ensure a good set of fruit, hand pollination should be carried out using a camel-hair brush to transfer the pollen from flower to flower.

When the fruit has set, the temperature may be increased a little and the atmosphere should be kept humid by damping down and syringing the plants again. Feeding with a proprietary liquid fertiliser should also be carried out until the berries show colour. After that, it is important to aim at a warm dry atmosphere in which the fruit will ripen well, although this does not mean that watering should be neglected. On the contrary, the strawberry in full growth will take copious supplies.

For large fruits the number on each plant should be reduced to about nine and the sooner this is done the better. Some form of support for the trusses of fruit will also be needed and this should be provided before the berries have a chance to flop over and become bruised against the side of the pot.

LETTUCES

The cultivation of good lettuces in greenhouses is not quite as easy as it may seem, and usually the amateur will get better results in the garden frame. In any case, to produce lettuces in winter demands a certain amount of heat and obviously this is used more economically in a frame, where it is provided preferably by electric soil-heating. Autumn and spring crops may, however, be taken from an unheated frame and as these are more likely to be attempted than a winter crop, I shall deal with them first.

For an autumn crop, seed should be sown in a seed tray about the end of August, using a cultivar such as Kwiek or May Queen. It should then be placed in the frame to germinate, as seedlings raised in the open do not do well when transferred to a frame. As soon as the young plants are large enough to handle, they should be pricked out into other trays at 1in apart. From then on they are grown in the frame, with ample ventilation, until they are about 2in high and ready for the final planting.

A soil in good 'heart', but not too rich, is required for these out-of-season lettuces and it is most important to make it firm, level and smooth on the surface. This initial preparation should be followed by a thorough soaking, after which the soil is allowed to become fairly dry on the surface before the plants are set out at 9in apart each way, care being taken to bury only the roots. After planting, the frame should be closed and left, with no further watering or ventilation except for an occasional airing after hard frost, until the lettuces are ready for cutting.

The spring crop follows much the same pattern, except that the seed is sown about the third week in October to provide plants for setting out in the frame in January.

If a heated frame or greenhouse is available, probably the best way by which the amateur can obtain a winter crop is by making small successional sowings from September to January and growing the plants on as described above. For this purpose, 'short-day' cultivars such as Cheshunt Early Giant and Kordaat should be used if hearted lettuces are required. A good alternative is Grand Rapids, which does not heart but produces a generous supply of tender curled leaves. A temperature of about 50° F (10° C) should be aimed at; and it is important to remove any dead or damaged leaves immediately, as otherwise grey mould (see Chapter 7) is sure to cause damage. Little watering should be needed, but the plants must never be allowed to dry out completely.

FRENCH BEANS

It is doubtful whether anybody nowadays forces French beans, but for those who are willing to provide the necessary temperature of about 65° F (18° C) in winter, they make a novel and useful crop. To provide a succession, seed is sown at intervals from August until February. Three or four seeds are set in a 6in pot of good soil, with room left at the top for a further top dressing. As the plants grow they will need supporting with twigs or canes; and as soon as the flowers are open, the plants must be kept in full sunlight and given ample water.

FORCED RHUBARB

This is one of the easiest of all crops to force, the main requirement being a good stock of mature outside 'crowns': once forced, these are of no further use. For a Christmas supply one or two strong crowns should be lifted in mid-November, care being taken not to damage the roots more than is necessary. If possible, they should then be left outside to become frozen before they are forced; but in any case by the end of November they should be placed under the staging in a heated greenhouse, where they should be well packed round with soil and kept in complete darkness. With constant moisture in a temperature of 60–70° F (16–21° C) they should then produce the first sticks about Christmas, but the best sticks are usually produced as a result of rather slower forcing.

To maintain a succession, further crowns should be brought in at intervals of two or three weeks.

VEGETABLE SEEDLINGS

To save time in the vegetable garden in spring and also to obtain rather earlier crops, several vegetables can be sown in seed trays and placed in gentle heat about the end of February. The resultant seedlings can then be pricked out and grown on in the greenhouse or frame until they are finally hardened off and planted out.

Cabbages, Brussels sprouts, cauliflowers, onions, leeks, lettuces and even shallots, set in boxes of soil from which they can be planted straight out, may all be treated in this way.

Chapter 11

Chrysanthemums

Although surrounded with some mystique springing from their success as exhibition flowers, by and large chrysanthemums are an easy crop as far as the production of good blooms for normal decorative purposes is concerned; otherwise, they would certainly not be as popular as they are among both amateur and commercial growers. Their only drawback is that they take up a lot of room in the greenhouse from October to Christmas.

Owing to the height of the plants, small greenhouses with permanent staging are unsuitable for chrysanthemums. Greenhouses of the glass-to-the-ground type, on the other hand, are ideal as they not only allow plenty of headroom for the plants but also admit the maximum light.

Artificial heat sufficient to keep frost out is very helpful, as it will go a long way towards keeping various troubles at bay. However, for those chrysanthemums that bloom in October and early November it is possible to do without it provided there is ample ventilation and no hard frost when the plants are in bloom.

Types of chrysanthemum

Although there are over a hundred species of chrysanthemum, the only plants we shall be concerned with here are the greenhouse ones of the sort that everyone knows and which constitute the most valuable of all autumn flowers. These chrysanthemums are divided largely into three groups: early, midseason and late, with the latter two mainly comprising the greenhouse ones. For exhibition purposes all three types are further divided into sections according to the National Chrysanthemum Society's classification. Of these the main ones among the greenhouse types are the Large Exhibition; Medium Exhibition; Exhibition Incurveds (large-flowered); Exhibition Incurveds (medium-flowered); Reflexed Decoratives (large-flowered); Reflexed Decoratives (medium-flowered); Anemone-flowered; Single-flowered; Pompons; and Sprays.

For those who intend to take up exhibiting, it is a good idea to join a local chrysanthemum society: the group will be only too pleased to welcome a new member and to give further details of the National Society's classification. The average grower is, however, more likely to be concerned with producing good blooms for decorative purposes only and the types commonly used for this purpose will therefore be the ones described here.

In catalogues these will be found under various headings such as Midseason Decoratives (October/November-flowering); Late Decoratives (December-flowering); Incurveds; Singles; and Sprays. The beginner is likely to be bewildered by the assortment of cultivars offered, but as these are constantly changing from year to year, there is little point in mentioning any of them here. Whichever he buys, the amateur is not likely to go far wrong if he follows the general points of cultivation given later in this chapter and takes note of any special cultural details with regard to stopping etc indicated in the supplier's catalogue.

Chrysanthemum terminology

As with all members of the botanical family Compositae to which the chrysanthemum belongs, what appears to be the bloom consists of a number of individual flowers or florets, each of which is complete in itself. In the single type of bloom these are of two kinds: the ray florets, which resemble petals and form the showy part of the complete flower-head; and the disc florets which form the flattish centre or 'eye'. Double flowers, which predominate among cultivated chrysanthemums, are composed almost entirely of ray florets, which may curve backwards to give a 'reflexed' bloom or inwards to form the almost globular 'incurved' one.

It is perhaps worth mentioning here that the ray florets are entirely female, while the disc ones have both male and female organs, the former being the stamens which provide the pollen necessary for fertilisation.

In the growing of chrysanthemums this fertilisation process is needed only to produce new cultivars by hybridisation; in ordinary practice the chrysanthemum is invariably raised annually from cuttings, which produce blooms in the same year as they are taken. But before their final flowering the plants pass through a number of stages, which can be controlled so that each plant produces a definite number of flowers at a definite time. It is largely in this control that the skill in growing for exhibition lies.

Left to itself a single-stemmed cutting grows upwards to about 10in to 18in high, when it forms a flower-bud which rarely develops properly. This is known as the 'break bud', because upward growth of the original stem then ceases and the plant breaks out into a number of leafy stems, which together form the natural or first breaks.

If the plant is still left to itself, each of these leafy stems produced in the axils of the leaves will in turn produce at its tip a flower-bud known as the 'first crown' bud. Further leafy shoots may then appear below this and these go on to produce the 'second crown' buds. Reference to 'terminal buds' is also made sometimes but this only confuses the issue. All that need be known about these is that when a cluster of buds appears at the end of a shoot, with no leafy shoots below it, the central main bud of the cluster is the 'terminal bud'.

In practice most chrysanthemums are flowered on either first or second crowns, but I shall deal with this in more detail later. In either case it is usual to disbud the plants by reducing the number of final buds on each stem to one, although if large individual blooms are not required several buds can be left to produce smaller flowers as 'spray'.

CULTIVATION

There are two main ways of growing chrysanthemums: one is to grow the plants in pots and the other is to plant them out in the garden in May for lifting and for replanting in the greenhouse in

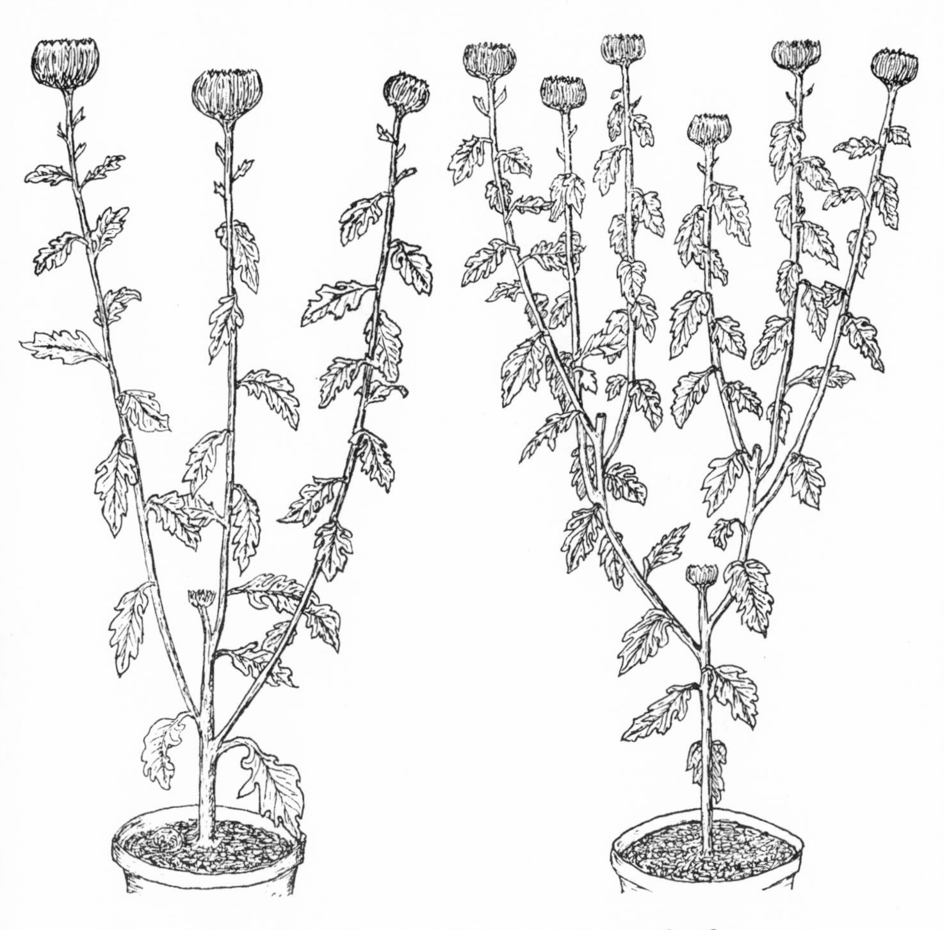

Fig 5 (left) a chrysanthemum plant stopped once for first crown blooms; (right) stopped twice for second crowns

the autumn. Obviously the latter is the more labour-saving way as there is no potting and much less watering to do, but not all cultivars respond well to this treatment and although the blooms on those which do are quite satisfactory for ordinary purposes, particularly when grown as spray, they are never as good as when the plants are grown well in pots. The December-flowering cultivars in particular do not do well as 'lifting' plants and few cultivars of any sort succeed when grown in this way on very light sandy soils.

Whichever method is followed, a start is usually made by purchasing rooted cuttings in spring, or as soon as they are available from the nursery. If these are intended for pot cultivation, first pot them up into 3in or 3½in pots, using the J.I.P2 and taking care not to pot them too firmly or too deeply. The same method may be used also for plants intended for lifting, but a simpler and more labour-saving way in this case is to make up in a cold frame a small well-drained bed of the same compost, about 4in deep, and set out the plants at 4in apart each way. The frame must be closed on frosty nights, but once the plants are growing well, plenty of light and air should be admitted whenever weather conditions allow.

After a week or two in the greenhouse, which should be heated only sufficiently to keep frost out, the young plants in pots may also be placed in a cold frame; but if they are kept in the greenhouse they must be given full light, with shade from only the hottest sun.

Potting on

It has long been the traditional method to pot the plants on a second time, from 3in pots to 4½in or 5in ones, before the final potting is done about the end of May, but the modern practice is to pot them up in early May from 3½in pots straight into the 8in, 9in or 10in ones in which they will flower. For the beginner the traditional method is probably better, as the young plants can be managed more easily in small rather than large pots, and there is also a considerable saving in space just when both the greenhouse and frame are normally filled to overflowing. The J.I.P2 is again suitable for these 4½in or 5in pots. The compost should be made slightly firmer than before, and the plants themselves should each be supported with a small cane after the potting is done.

Cool and airy conditions, either in the greenhouse or in the frame, should then ensure strong steady growth until the plants have filled their pots with roots and are ready for the final potting, when the J.I.P3 should be used over a good layer of crocks covered with coarse peat to ensure perfect drainage. This time the potting compost should be made quite firm without ramming it unduly and a space of at least 1in should be left at the top of the pot for watering. As before, the plants should not be potted deeply and the top of the soil-ball taken from the previous pot should finish just below the surface. For most cultivars a 9in pot should be used for each plant: although weak growers may need only 8in and strong ones 10in, this is something which can be learned only from experience, and to start with it is best to use 9in pots for all the plants.

Standing out

After potting, the plants will have to be put out on a standing ground. Ideally, this should be a well-drained flat area covered with gravel or weathered ashes, but any level surface will do if the pots are stood on slates or pieces of flat asbestos sheeting. To allow for easy working a path at least 30in wide should be made between each two rows of the plants, which should be spaced at 18in apart each way, centre to centre, in the double row.

It is often recommended that the standing ground should be in a position sheltered as much as possible from the wind, but such a situation can often be worse than a completely open one owing to the eddies and 'whirlpools' of wind that are likely to be created there. A better method is to have the plants in the open and to screen them round with small-mesh wire netting, which by filtering the wind reduces its force without diverting it.

Before the plants are placed on the standing ground, some provision must be made for supporting them so that they can be tied securely as soon as they are in position. For this purpose a length of stout wire should be strained between two strong end posts, each supported on the inner side by a diagonal stay. The height of the wire will depend on the eventual height of the plants, but a useful height is 4ft as this will allow either 4ft or 5ft canes to be used. At a pinch one cane will do for one plant but it is better to use three canes to each, tying the two outer canes to the wire and letting the third one stand free on the inside of the double row, where it will be reasonably protected from the wind. Each double row will need two wires, of course.

The tying of plants to canes with a soft horticultural twine must be done as soon as they are in position. The correct way to do this is to tie the twine securely to the cane so that it cannot slip down and then tie it again in a loose loop round the stem. Further tying must then be done as soon as the stems have grown sufficiently to move freely again.

Watering and feeding

Care in watering is essential right from the start and, up to the final potting, the best method is to allow the compost to dry on the surface then to soak it thoroughly. When the plants are well established in their final pots, however, copious watering will be needed, particularly on hot days, when it may be necessary to water the plants two or three times a day. If a hose can be used conveniently, this should have its open end covered with a cloth to prevent the water from gushing into each pot; otherwise a watering-can will have to be used to provide enough water to soak each pot right through to the bottom. To reduce the amount of watering needed, the surface of the pots may be covered loosely with potting compost or some form of humus, even grass mowings as long as they are applied loosely.

Feeding

This should be started four or five weeks after the plants were placed on the standing ground, or as soon as they are obviously growing well. A proprietary chrysanthemum fertiliser should then be used strictly according to the maker's instructions, but always make sure that the soil is moist before the fertiliser is applied. The feeding can be continued right through the summer until the flower-buds show colour, when it should be stopped.

Lifted plants

The rooted cuttings which were planted in a frame for setting out later in the open ground should be ready for their summer quarters by the beginning of May, or a week or two later in the north. The ground for them should have been well prepared beforehand, by digging it thoroughly and working in plenty of garden compost or old manure, together with a dressing of a complete chrysanthemum fertiliser. A mixture of 5 parts sulphate of ammonia, 8 parts superphosphate of lime and 2 parts sulphate of potash (parts by weight) may be used instead of the complete fertiliser. If there is a chance that the soil might be acid, it is a wise precaution to use a soil-testing outfit and if necessary add lime to bring the pH up to 6.

Before planting, the ground should be made firm and level. The plants should then be set out in double rows, with 18in each way from plant to plant and with a 3ft path between each double row and the next. Firm planting, without burying the base of the stem, is essential and the initial watering in will be simplified if each plant is set in a slight depression in the soil.

Staking is done by providing each plant with one 4ft or 5ft cane inserted well into the ground, the stems being looped to this with soft twine tied to the cane so that it cannot slip down.

Feeding and other subsequent operations are then carried out as to pot-grown plants.

STOPPING AND DISBUDDING

The natural growth of the chrysanthemum has been described already and even if left to itself the plant will produce reasonably acceptable blooms in its own good time. It is more usual, however, to adapt its growth by stopping and disbudding to control the time of flowering and the size and number of blooms. For exhibition purposes these operations demand a high degree of skill and careful timing, but for good decorative blooms such as you find in a florist's, the following stopping programme is all that should be needed.

Cultivars for October/November flowering

Allow the plants to break naturally, without stopping. If they have not done so by the last week in June (or mid-June in the north) pinch out the tip of the stem. With this treatment the plants should each produce six to ten blooms on first crowns, but if fewer stems than this are produced, two or three of the uppermost ones should be stopped at the first pair of leaves.

Singles for November flowering

Stop in early May and again in early July.

December-flowering cultivars

Stop in early May, retain the best three or four of the resultant stems and stop these again in the last week of June for second crowns; or stop once only for first crowns in the last week of June. (These late-flowering cultivars vary considerably in their stopping requirements as far as flowering at Christmas is concerned, and to get the best results from them is largely a matter of experience.)

If any of these methods of stopping should result in the buds forming too early, flowering can be delayed by 'running them on', which means that the early buds are removed completely together with all the leafy sideshoots around them except the topmost one. This one is allowed to grow on and flower.

Whichever system of stopping is practised, the job should never be done within ten days either way of any planting or potting. It is important also to remove no more than the very tip of the stem, back to the topmost pair of leaves. There is much to be said for growing the same cultivars for at least a few years, particularly in the case of the Christmas-flowering ones, since if a record is then kept of the stopping, flowering and disbudding times it should soon be possible almost to guarantee a good crop of blooms just when you want them.

Disbudding

By August or September all the plants should have produced the buds which are to mature into blooms. If large blooms are wanted disbudding must be carried out. This simply means retaining or 'securing' the best bud of the cluster which forms, usually the centre one. All other buds must be removed by breaking them off gently with the tip of the thumb as soon as they are large enough for this to be done without risk to the main bud. A job requiring considerable care, it is best done by holding the main bud in one hand while the other buds are removed.

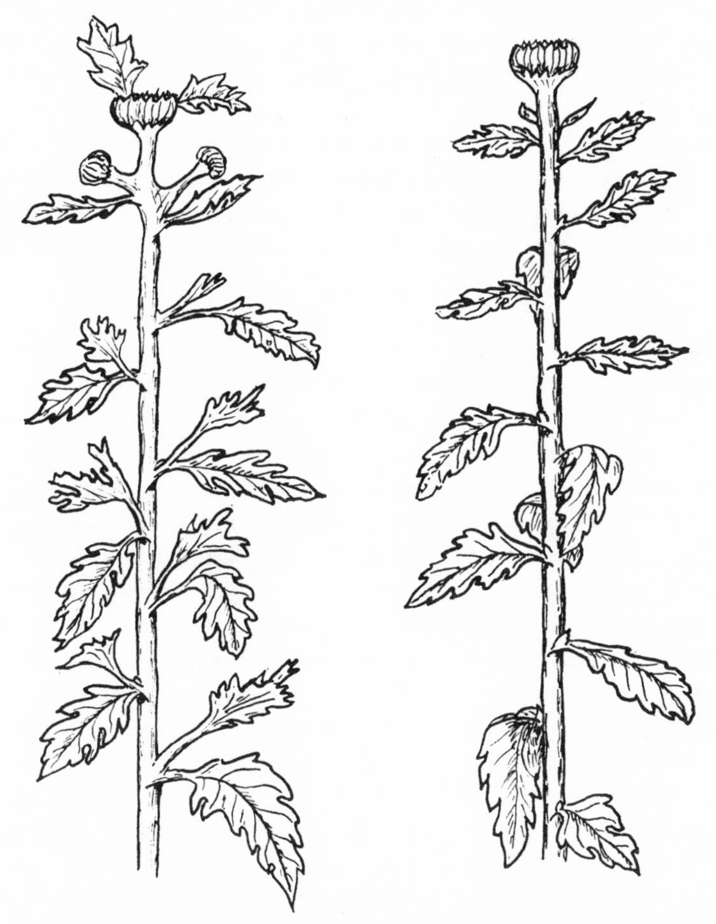

Fig 6 Before and after disbudding: all the sideshoots and unwanted flower-buds are removed as shown on the right

All leafy shoots below the bud must be removed also, by breaking them out gently in a downward direction.

Housing the plants

This takes place towards the end of September, when most of

the final buds have formed already even if they have not started to show colour. Disbudding may if necessary be continued after the plants have been brought inside.

Before bringing the plants in, make sure that the greenhouse has been cleaned thoroughly, including the glass, and give the plants a good spraying with an insecticide and fungicide. Damaged and dead leaves should be removed, together with any weeds which have appeared in the pots. Plants grown in the open ground for lifting should be treated similarly but in addition each plant should be cut round with a sharp spade a week to ten days before housing.

When carrying the plants through the doorway, always put the pot or root through first as otherwise there is a risk of the stems being broken off against the doorposts. Then space them out so that there is a free circulation of air around each one and adequate access to them.

For pot-grown plants maximum ventilation should be given for the first few days, with the door and ventilators fully open except when there is a risk of frost. Lifted plants need the opposite treatment in that they should be kept 'close', with only the minimum ventilation, until they have settled down, when more air should be given gradually. These lifted plants will be grown in the actual soil of the greenhouse and the simplest way to plant them is to take out a trench, stand the plants in it and cover their roots very firmly with the excavated soil, which should be well watered in around them. Once they are established in the greenhouse both pot-grown and lifted plants need an airy buoyant atmosphere, with preferably enough heat to keep the air on the move. Waterings must be done carefully at all times—do not splash the water around more than necessary—and all dead or damaged leaves should be removed immediately.

PROPAGATION

As soon as the flowers are over, the plants must be cut back to about 8in high and kept quite cool and only just moist. From these cut-back plants or 'stools', the cuttings which will provide the next year's crop are raised, usually in February and March for those cultivars intended for cut bloom, although some exhibition plants are rooted earlier than this.

Traditionally, shoots springing from the actual base of the plant have always been regarded as the most suitable for use as cuttings; but modern research has shown that those taken from the stem will do just as well, and this can be a great help with some cultivars which are very reluctant to produce basal growths. The actual type of cutting is, it seems, more important then where it comes from. The ideal cutting should be 2in to 2½in long, short-jointed and just over ⅛in thick.

For rooting the cuttings, the J.I.P1 is an excellent medium. This should be made firm and level in a seed tray in the same way as when one is preparing boxes for seed-sowing. Some commercial growers do no more than snap the cutting off the plant and push it into the compost just as it is, but it is advisable to be rather more fussy. Trim off each cutting beneath a joint with a sharp knife and remove the lower leaves before inserting it into the compost, no more than 1in deep. About 1½in each way is a suitable spacing.

After insertion the cuttings should be given a thorough watering and stood on the greenhouse bench, when they will need to be shaded from full sun. No further water should be given until the cuttings show signs of flagging from dryness, when they should be soaked thoroughly again. In a temperature of 45–50° F (7–10° C) this should be sufficient to carry them through until they have rooted and are obviously growing, usually in about three weeks. They are then ready for potting up as described earlier.

ALL-THE-YEAR-ROUND CHRYSANTHEMUMS

The attractive pots of dwarf chrysanthemums which now appear in the shops all the year round must certainly arouse the curiosity of many amateurs who would probably like to grow some. It must, however, be said here that this is hardly a job for the amateur: apart from the fact that many of the cultivars used are seldom included in catalogues, the whole procedure is a complicated one.

Very briefly, the production of these plants is based upon photoperiodism or the relation of the flowering period of a plant to the length of day. Many plants form their flower-buds only when there is a certain number of daylight hours and thus they may be either long-day or short-day plants. The chrysanthemum, being naturally an autumn- and winter-flowering plant, obviously comes into the latter class.

To induce the plants to flower earlier than they normally would, the day is shortened artificially by covering them with black cloth for a certain time each day over a definite period. Conversely, if flowering is to be delayed, the plants are illuminated artificially to give the effect of increased day length.

With careful treatment in this and other respects the plants can be brought into bloom at any given time while they are still small, but after flowering they will revert under normal conditions to their original tall habit.

Page 105 (*above*) Single cuttings of many plants can readily be rooted indoors beneath an inverted jar; (*left*) the different humidity needs of tomatoes and cucumbers can be met by screening the cucumber area off with polythene sheeting

Page 106 (*above*) Tomatoes being grown on the ring-culture system; (*below*) a modified form of ring culture, with the plant containers half-buried in the aggregate to reduce watering

Chapter 12

Other Cut Flowers

To produce good carnations, particularly in winter, is indeed a test of the grower's skill; but this should not put anyone off growing them, since with normal care quite good results can be expected provided the greenhouse is very light and airy and warm enough to keep frost out.

The only type of carnation grown to any extent under glass is the one known as Perpetual Flowering: a rather misleading term, as it does not mean that the plants are constantly in bloom but that they can be brought into flower at any time of the year. As with chrysanthemums and other plants, new varieties of carnation are being introduced every year, but many well-tried ones still persist. The following are a few which are particularly suitable for the amateur grower: Northland and White Sim, both white; Allwood's Prolific and Ashington Pink, both pink; Spectrum and William Sim, scarlet; Royal Crimson and Topsy, crimson; Royal Lavender and the strongly scented Doris Allwood, salmon pink.

Young plants from 3in pots are usually available from April to June. They are normally sent out already stopped and ready for moving on into larger pots.

Potting and summer management

Either 6in or 8in pots may be used, the latter enabling the plants to be grown on for three years instead of for two with 6in ones. Clean well-crocked pots are essential, and a suitable compost is the J.I.P3 in which the plants should be made as firm as possible, with the top of the old soil-ball about ¾in below the rim. One good watering in should be enough to start them into growth again, no more water being given until the compost is almost dry.

After potting, the plants may be stood outside for the summer, but if there is space available they will do far better and be managed more easily on the greenhouse bench. Here they should be kept rather close, with a minimum of ventilation, for the first few days. It will be helpful also if they are syringed lightly overhead, although this should not be overdone once the plants are growing away strongly, as carnations resent a humid atmosphere.

Watering must be carefully attended to: although the plants do not need a lot of water, they must never be allowed to dry out. A fairly safe procedure in summer is to wait until the compost is quite dry on the surface and then to soak the soil thoroughly. In winter it is best to let the compost practically dry out between soakings.

Maximum light and air must be given at all times once the plants are established, and during hot weather a limited amount of syringing and damping down can be carried out. These operations should cease by the autumn, when the problem is more likely to be how to keep the air dry enough rather than to prevent it from becoming too arid.

Stopping and disbudding

About a month after potting, the shoots from the original stopping will themselves need stopping. Each shoot should be pinched back to six pairs of leaves. On a few varieties which break naturally this may not be necessary, but as a general rule if there is no sign of sideshoots forming, the stems should be pinched out a few at a time. To know when exactly to do this requires a little experience, but the idea in general is to stop those shoots that are neither too soft and green nor too hard and woody.

The time of this stopping also affects the time of flowering, which takes place usually about five months after the shoots have been stopped, provided adequate warmth is available. By stopping only a few shoots at a time, you ensure that the plant receives less of a shock and that the flowering period will be extended. Much, however, depends on whether or not the greenhouse is heated. In a warm greenhouse stopping may be carried out safely up to the end of July to provide autumn and winter flowers, but in a cold house mid-June is the latest date for ensuring autumn blooms. For spring ones an August stopping will give good results. It is not a practical proposition to attempt winter flowers without heat.

For large blooms the plants will have to be disbudded by the removal of all flower-buds springing from the axils of the leaves, so that only the main one at the top of the stem is left. The sooner these unwanted growths can be pinched out the better, but any leafy sideshoots on the lower part of the stem should be left intact.

Staking, feeding, etc

Potted in a good compost the plants should not need feeding during the first summer until about August, when a proprietary carnation fertiliser should be used as recommended by the makers. Alternatively, a suitable homemade mixture is 2 parts sulphate of ammonia, 3 parts superphosphate of lime and 1 part sulphate of potash (parts by weight), used at the rate of half a teaspoonful per plant each fortnight. The compost should be moist before any fertiliser is applied. No feeding should be carried out after the end of October, but in the case of plants which are being carried on for a second or third year, it should be started again in February.

As the plants grow quite tall—a few will even reach 6ft in their second year—some form of support will be needed. This is best provided by means of the special wire carnation rings that are available for use with a cane inserted in each pot. These should be placed in position at an early stage before there is any tendency for the plants to flop over, and preferably as soon as they have been potted.

The cutting of the flowers is usually all the pruning that is necessary, provided they are taken with reasonably long stems. After that the plants are grown on as before for a second year.

Propagation

Perpetual carnations are propagated by cuttings taken between mid-November and mid-February. This is a job which requires infinite care if a clean healthy stock is to be maintained. Ideally, special stock plants, which are allowed to retain their flowers only until the best plants have been selected, are grown to supply the cuttings. For most amateurs, however, this will be impracticable and the cuttings may be taken instead from the most healthy and free-flowering first-year plants.

The best shoots for cuttings come from the lower half of the plant and consist of sideshoots about 4in long, with several leaves close together at the base. When severing shoots from the parent plant use a sharp knife. Do not take them right back to the main stem but cut them immediately beneath a joint so that a portion of the shoot is left attached to the main stem; this remaining portion will then produce more shoots. The only further preparation that is needed is the removal of one or two pairs of leaves from the base of the shoot.

For small numbers of cuttings it will be found most convenient to insert them in 5in pots, which should be perfectly clean and well crocked. A layer of coarse peat should then be placed over the crocks before filling the pot almost to the rim with clean sharp sand or vermiculite, which must be made quite firm and level. These preparations should be carried out before the cuttings are taken, so that there is no delay between the severing of the shoots and their insertion. About 1in apart is a suitable spacing for them and on no account must they be inserted deeply, certainly no more than ¾in.

After insertion give them a thorough but gentle watering with a fine-rosed can and place them in a propagating frame, which should be kept closed for the first twenty-four hours. Then give just enough ventilation to air the frame daily, increasing this gradually until the cuttings show signs of growing, when the cover of the frame should be left off.

An air temperature of 45–50° F (7–10° C) is about right for the cuttings, although they will appreciate a slightly higher bottom heat. Kept just moist in this, by being watered only when the sand shows signs of drying out, they should root in about a month, when they should be lifted carefully and potted into 2in pots, using the J.I.P1.

In this and subsequent pottings only the roots should be covered; the stem should be left fully exposed.

The plants can stay in these small pots for about a month, when they will need potting on into 3in ones, again using the J.I.P1. Finally, when they have made about six pairs of leaves, they should be stopped and grown on in the same way as the original purchased plants.

FREESIAS

Freesias are a relatively easy crop to grow either as pot plants or cut flowers. For the latter purpose they are grown commercially in beds or borders under glass, but for the amateur either pots or boxes provide a simpler method, boxes being most suitable if the flowers are required in any quantity.

Freesias can be grown from both seed and corms, but the latter need only half the time and are considerably easier to manage as they do not have to be kept on the move all summer. Corms planted in August and September will flower in spring, provided a frostproof greenhouse is available and no attempt is made to bring them along earlier by giving them more warmth: this will result only in a lush growth of foliage and no flowers.

For pot plants, five or six corms are generally grown in a 5in pot, but if boxes are used these should be from 4in to 6in deep, with the corms planted about 2in apart, centre to centre. The J.I.P2, over good drainage, is an excellent compost to use but although freesias are a little touchy about the type of soil they like, they will usually do well in any reasonably good one with plenty of peat or leaf-mould added, for they seem to prefer a high humus content. Whether pots or boxes are used, the corms should be covered with about 1in of soil.

When they have been potted or boxed, the place best for the plants is in a cold frame where, if the compost was reasonably moist to start with, little or no water will be needed until growth starts after a few weeks. To prevent the compost from drying out, the pots or boxes can be covered with a layer of moist peat and the frame shaded in sunny weather; then if it still tends to dry out too much, only enough water should be given to barely moisten it.

During the time the plants are in the frame, the frame-light must be kept on but open enough to admit plenty of air. As soon as the corms have sent up shoots about 1in high, the plants should be removed to a frostproof greenhouse, where they must be kept well up to the glass to ensure sturdy growth. Gradually more water will be needed as the plants grow and staking with twigs or with canes and twine must be done early, well before the foliage tends to flop over.

After flowering, the corms can be kept for another year by feeding and watering them well until the foliage yellows and dies down, when the pots or boxes should be placed in full sun in the greenhouse so that the corms are fully ripened. During this ripening period no water is needed and in August or September the corms can be shaken out of the old soil and repotted.

Cultivation from seed

Freesia seedlings do not transplant well, so the seed must be sown in the pots or boxes in which the plants are to flower. Plants raised from seed sown from April to early June will normally flower during the period September to spring.

To ensure a satisfactory crop of seedlings, the seed should be mixed with moist peat and kept in a closed jar in a temperature of about 70° F (21° C) for three or four days, until it has sprouted. Each sprouted seed should then be sown in the pot or box at the same spacing as for corms but only about ½in deep. However, as there may be some failures, it does no harm to sow a few extra seeds and remove any overcrowded ones later. Warmth, darkness and moisture are needed to complete the germination, but after that the plants should be grown in as cool and airy conditions as possible, say in a cold frame fitted with a slatted light in sunny weather. The only time when the glazed light will be needed is during periods of heavy rain. By September, the plants can be brought inside and treated in the same way as those grown from corms.

GERBERAS

Among the less common greenhouse cut flowers is *Gerbera jamesonii*, the Transvaal or Barberton Daisy, which in its modern hybrid forms produces magnificent daisy-like blooms, usually single and up to 5in across, in a wide range of delicate pastel shades. Carried on stems up to 18in long these are superb flowers for cutting, and although in a cold greenhouse they will be produced only in summer their season can be prolonged well into the winter with the help of a little heat, say 45–50° F (7–10° C).

For their maximum annual yield of thirty to forty blooms per plant, gerberas need plenty of room for their large fleshy roots and are best planted therefore in the actual soil of the greenhouse, where they can remain for two or three years. A more practical proposition for the amateur, however, is to grow each plant in a 10in pot or three or four plants in a large tub so that, if necessary, they can be placed outside for the summer.

Any moderately rich firm soil suits the gerbera provided this is well drained, for one of the great drawbacks with this plant is its tendency to rot off in a wet soil in winter. To guard against this the plants must be put in with their crowns well above the surface. A good method is that of filling the pot or container with soil only up to about 2in from the rim, covering this with about 1½in of shingle and finally setting the plant in position so that the base or crown is just above this. Maximum light and plenty of watering and feeding are needed in summer but in winter the plants must be kept much drier and feeding is not required.

Propagation may be effected by the division of old plants in spring or by seed. The seed must be sown as soon as it is received, since it deteriorates rapidly. March or April is the best time for sowing and the fairly large flask-shaped seed should be sown preferably in an upright position, with the point downwards and the tufted top about ¼in beneath the surface of either the J.I. seed compost or a soilless one. A temperature of at least 70° F (21° C) should then ensure fairly quick germination. As soon as the seedlings are large enough to handle, they should be potted off into 3in pots and grown on until they are large enough for their final quarters.

NARCISSI AND TULIPS

Many of the popular narcissi, including daffodils, and also the taller tulips such as the Darwins, can be grown successfully as cut flowers in the greenhouse, although not all of them will respond to forcing. Where the aim is merely to produce flowers earlier than those grown outside, there should be no difficulty in getting good results. Narcissi Paper-White and Soleil d'Or are particularly useful for providing early bloom with only the minimum of heat, while one of the easiest Darwin tulips for bringing along early, although needing more heat than the narcissi, is the pink Rose Copeland.

For cut flowers, the usual method is to grow the bulbs of both narcissi and tulips in boxes about 4in deep, filled with a good open soil in which the bulbs are planted with their 'noses' just above the surface. A space of 1in or so between the bulbs is about right and they should be planted just firmly enough to prevent too much compacting of the soil beneath them. Acid soil, incidentally, should not be used for tulips and if there is any doubt about this a little lime should be added.

The bulbs of daffodils and other narcissi may be boxed in succession from August to October and those of tulips during September and October. The bulbs should then be ready for taking into the greenhouse two or three months later, but the actual time will depend on the condition of the bulbs, which must have made plenty of root growth and produced shoots an inch or two long. If there is any doubt as to whether they are ready for the greenhouse, it is advisable to give them another week or two, for if they are brought in too early they may be a failure.

When the boxes have been planted they should be placed outside and covered with about 4in of sand or weathered ashes: they will need no further attention until they are brought inside. For the first few days after this, only a little heat should be given, but this may then be increased gradually to bring the plants into bloom earlier.

Chapter 13

Summer Bedding Plants

One of the great advantages of a greenhouse, particularly a heated one, is that it enables the gardener to grow his own bedding plants for summer and if for no other reason it is certainly worth growing these just to avoid the considerable expense of buying them. It often also enables the gardener to have better plants than those which are normally available in the shops, for these have often suffered at least some ill-treatment after they have left the nursery. What is more, a much wider choice is possible as far as the cultivars of the different plants are concerned.

Plants from seed

Many of the most popular bedding plants are grown from seed, but there is no need to discuss the actual sowing and pricking out of them here as these operations have been dealt with already. More important at this stage are the times of sowing, for these largely decide the times when the resultant plants will be ready for the garden.

Apart from antirrhinums, which may be planted out towards the end of April if they have been well hardened off first, the summer bedding plants can seldom be planted safely until well into May, and some not even until June. The sowings must therefore be timed to fit in reasonably with these planting dates: if they are sown too early, the plants will become starved and crowded before they can go out; if they are sown late, the plants will be equally late to flower.

It is impossible to give any precise times of sowing as these can vary in different parts of the country and also according to the weather; a week or two either way will not make a great deal of difference. As a general guide, the times given in the following list should give satisfactory results and if a simple record is kept, the sowings and plantings can be amended to suit individual conditions.

The times given are for heated greenhouses. With cold ones it is usually wisest to wait until the end of March before making any of these sowings other than antirrhinums. If a warm propagator can be used indoors, a start with all the plants mentioned can be made in mid-March, so that by the time the plants need pricking out they will be reasonably safe even without heat in the greenhouse.

SUMMER BEDDING PLANTS FROM SEED

African Marigold see *Tagetes erecta*

Ageratum mexicanum Floss Flower. Sow early March, temp 55–60° F (13–16° C). Plant out late May. A useful dwarf plant for bedding, edging and window boxes; also for pot cultivation in the greenhouse. Pinch out at 3in high to ensure busy plants.

Alyssum maritimum (syn *Lobularia maritima*) Sweet Alyssum. Sow end of March, temp 45–50° F (7–10° C). Plant out mid-May Excellent for edging and window boxes.

Amaranthus tricolor Joseph's Coat. Sow end of March, temp 55–60° F (13–16° C). Plant out early June. 18in high, grown for its brilliantly coloured foliage.

Antirrhinum Snapdragon. Sow early February, temp 45–50° F (7–10° C). Plant out late April. Excellent for formal bedding.

Aster (China) see *Callistephus*

Balsam see *Impatiens*

Begonia semperflorens Sow early February, temp 60° F (16° C). Prick out and grow in 3in pots for planting out in early June. 12in high, good for formal bedding, pots and window boxes.

Callistephus China Aster. Sow mid-March, temp 50–55° F (10–13° C). Plant out mid-May. Excellent for cutting as well as bedding, in both its single and double forms.

Canna see Chapter 16 on plants from bulbs, pp 155–68.

Cosmos (*Cosmea*) Sow mid-March, temp 55–60° F (13–16° C). Plant out end of May. At about 3ft high better for cutting than bedding.

Dahlia Sow the bedding types early April, temp 50–55° F (10–13° C). Plant out early June from 3½in pots.

Dimorphotheca aurantiaca Namaqualand Daisy or Star of the Veldt. Sow mid-March, temp 50–55° F (10–13° C). Plant out end of May in light soil in full sun. 18in.

Felicia bergeriana Kingfisher Daisy. Sow mid-March, temp 50–55° F (10–13° C). Plant out mid-May. 6in. Good for edging, bedding and window boxes.

French Marigold see *Tagetes patula*

Impatiens Balsam and Busy Lizzie. *I. balsamina*, the balsam, grows easily from seed sown in early April, temp 50–55° F (10–13° C), but it is so fragile that it is better grown as a pot plant for the greenhouse, a purpose for which it is ideal. The Busy Lizzie (*I. holstii* and *I. sultani*), although usually grown as a pot plant, also makes a good bedding one, either from cuttings or from seed sown in early February, temp 60–65° F (16–18° C). Plant out early in June.

Kochia tricophila Summer Cypress. Sow mid-March, temp 45–50° F (7–10° C) and plant out early June. Grown for its finely cut foliage, green in summer, changing to crimson later. 2–3ft.

Lobelia erinus Sow end of February, temp 55–60° F (13–16° C). Plant out mid-May. A popular edging plant about 6in high.

Matthiola Ten-weeks Stock. Sow mid-March, temp 50–55° F (10–13° C). Plant out mid-May, in a well-drained sunny soil with lime. Only the double-flowered forms are worth growing, but most seed strains give a proportion of singles. With the one known as Hanson's Double, it is possible to select the doubles at an early stage by discarding all seedlings with dark green leaves.

Mesembryanthemum criniflorum Livingstone Daisy. Sow early March, temp 50–55° F (10–13° C) and plant out mid-May. Invaluable for providing a low carpet of vividly coloured 'daisies' in full sun. Suitable for edging, bedding and rock gardens.

Nemesia Sow late March, temp 50–55° F (10–13° C), for planting out mid-May. The dwarf compact varieties are the best ones to grow and these are usually indicated in catalogues.

Nicotiana affinis (syn *N. grandiflora*) Tobacco Plant. Sow early April, temp 50–55° F (10–13° C) for planting out end of May. Harden off well before planting as the large leaves are easily damaged by cold winds. The old *N. affinis*, with very fragrant white flowers opening in the evening, has now been ousted largely by more modern forms that stay open all day, and for fragrance the white varieties of these should be grown.

Petunia Sow in mid-March, temp 50–55° F (10–13° C), and plant out mid-May in not too rich soil, in full sun. Nowadays there is a multitude of varieties of this extremely colourful plant, among them both single and double forms that are useful for window boxes or as greenhouse pot plants, as well as for bedding.

Phlox drummondii Sow mid-March, temp 50–55° F (10–13° C). Plant out mid-May and peg down the trailing stems as they grow.

Portulaca grandiflora Sow mid-March, temp 55–60° F (13–16° C). Plant out mid-May as an edging plant or in the rock garden. Full sun and well-drained soil essential.

Salpiglossis sinuata Sow mid-March, temp 50–55° F (10–13° C) and plant out mid-May. At 2ft to 3ft high, a better plant for mixed borders than for bedding.

Salvia splendens Sow early February, temp 60–65° F (16–18° C). Prick out and then pot up into 3½in pots of rich soil. Keep warm and moist until early June, then plant out in rich soil in full sun.

Sedum caeruleum Sow mid-March, temp 50–55° F (10–13° C), and plant out mid-May. A little-known dwarf plant with blue flowers, it makes an excellent subject for edging or for the rock garden.

Stocks see *Matthiola*

Tagestes erecta African Marigold. Sow mid-March, temp 50–55° F (10–13° C) and plant out mid-May. Does well in partial shade. The dwarf varieties are most suitable for bedding, the taller ones for cutting.

Tagetes patula French Marigold. Sow and plant out as for *T. erecta*. Useful for bedding and window boxes in the shade as well as in the sun.

Tagetes tenuifolia pumila (syn *T. signata pumila*) As for *T. patula*. A more dwarf plant suitable for edging.

Tobacco Plants see *Nicotiana*

Verbena hybrids Sow early February, temp 60–65° F (16–18° C). For the best results prick out the seedlings, then pot them into 3½in pots for planting out late May. Rich well-drained soil and full sun. The spreading growths may be pegged down to form a carpet around taller plants.

Zinnia This is one of the most tender of all the half-hardy annuals. As it grows quickly, it should not be sown before mid-April, temp 50–60° F (10–16° C) for planting out early June. A rich, moist, warm soil in full sun is essential and successional sowings may be made in the open until the end of May. There are dwarf varieties for bedding and taller ones for cutting.

BEDDING PLANTS FROM CUTTINGS

All the bedding plants mentioned above are treated as half-hardy annuals, though in some cases they are, strictly speaking, perennial. There is, however, another group of plants that can be grown on from year to year: the usual practice with these is merely to keep the old plants as a source of cuttings which, taken each year, make more vigorous and floriferous plants than the old ones.

The only drawback with these plants from cuttings is that, unlike seedlings, they can transmit any disease inherent in the parent plant. For this reason it is essential to use only the best and healthiest plants for propagation purposes, and to 'rogue' out any doubtful ones during the summer so that these are not propagated for the following year. This is particularly important with zonal pelargoniums (geraniums) as these suffer from several diseases, one of which can be very deceptive. This is the virus disease 'leaf curl' that results in spotted crinkled leaves which turn brown and shrivel up, so that eventually the whole plant looks anything but healthy. Later in the summer the plants appear to recover, but the disease is still there and any cuttings taken from affected plants will have the same trouble the following year.

The bedding plants grown from cuttings can be divided roughly into two groups: those which need heat to keep them alive in winter and those which are quite safe in cold frames. The appropriate method for each plant is indicated in the following list, but as the main principles of propagation by cuttings have been dealt with already in Chapter 4, additional details are given only where necessary.

Calceolarias These should not be confused with the very showy herbaceous calceolarias grown in summer in the greenhouse, for they have much smaller flowers, yellow or bronze, on dwarf erect plants. Cuttings of the unflowered sideshoots should be taken 2–3in long in September or October and inserted in sandy soil in a cold frame. Give no ventilation at first but a little more when the plants are obviously growing. If the cuttings are spaced 4in apart each way in the frame, they can be left there until mid-May, when they can be hardened off and planted out. If spaced closer than this, they should be potted up in March and grown on in the frame until they are planted. To ensure that the cuttings are free from greenfly, swish them round in a bucket of insecticide before inserting them.

Centaurea The two species used for bedding, *C. argentea* and *C. ragusina*, are both grown for their fern-like silvery leaves, which make an excellent foil for other plants in flower. Neither of them is hardy and cuttings should be taken in August or early September, using short unflowered side shoot staken with a heel. After their lower leaves have been trimmed off to give about 1in of bare stem, the cuttings are inserted singly in small pots of light compost and kept in a frostproof frame or greenhouse through the winter, with very little water but plenty of air. Eventually they are transferred to 3½in pots and planted out finally in early June.

Chrysanthemum frutescens Better known as the marguerite, this plant is not as popular as it used to be: once a stock has been obtained it pays to look after it, as it is not always easy to get more. In early October lift some of the best plants and pot them, then keep them safe from frost and fairly dry in the greenhouse. In early spring cut them back and give more warmth and water to induce the growth of new shoots, which can be taken off 2–3in long and rooted. Pot them up eventually and grow on until planting takes place in early June.

Fuchsias Lift and pot the plants in October, then keep them safe from frost and almost dry until early February when they should be cut back to about 6in high. For further details see the list of popular plants in Chapter 19, pp 184–95.

Gazania splendens In mild areas, 2in long unflowered cuttings of this Treasure Flower may be safely inserted in a cold frame but elsewhere cool greenhouse protection is needed. Root the cuttings in seed trays in August or September and bring them into the greenhouse, where they should be kept safe from frost and almost dry through the winter. Pot them up in spring and plant out in early June.

Heliotropium This once popular Cherry Pie is not as common as it used to be, partly no doubt because it needs a good deal of heat to bring it safely through the winter. If plants are available, take unflowered sideshoots in August or September and root these in pots or boxes, after which they must be kept at a temperature of at least 50° F (10° C) through the winter. In spring, start the cuttings into growth and use the new shoots from them as further cuttings, for these will do better than the original ones. Few bedding plants are more tender than the heliotrope and for this reason planting out should not take place until mid-June. Cuttings rooted in spring may also be grown on in the greenhouse to make attractive and fragrant pot plants.

Iresine herbstii This Blood Leaf plant is grown only for its deep red leaves and usually it is not allowed to bloom. Lift a few plants in September or early October and pot them up for storing in a frostproof greenhouse. In February cut them back lightly and use the young shoots as cuttings, which should be pinched out at an early stage to ensure bushy plants. Plant out in early June.

Lantana camara As well as making an attractive greenhouse plant for summer this small shrubby plant, normally available in hybrid forms, is also useful for planting out in early June for bedding. Cuttings of the unflowered sideshoots, taken about 2in long, will root readily in a warm frame in the greenhouse in August or September and these should then be kept warm and fairly dry through the winter. Pot them up in spring and grow on with plenty of light, warmth and water, plus generous feeding when the plants have filled their pots. Cuttings may also be taken in spring, in a close propagating frame at 65–70° F (18–21° C), and seed too may be sown in a similar or preferably higher temperature.

Pelargoniums Cuttings of the zonal pelargonium (better known as the geranium) may be rooted in late summer or in spring, but the former is definitely the easier time. In August, for instance, half-ripe unflowered shoots (those which are just turning red in colour) will root readily in the open ground if inserted firmly in sandy soil in full sun. About 4in long is a suitable length for the cuttings, which must be potted up and placed in a frostproof greenhouse as soon as they have rooted. Similar cuttings may also be taken in a warm greenhouse in the autumn, and then if the old plants are placed close together

Page 123 Well-trained grape-vines with mature 'rods' or stems carrying a fine crop of fruit

Page 124 (*above*) When potting up cultivars of chrysanthemums it is essential to label each one as it is dealt with; (*below*) when stopping chrysanthemums, the tip of the stem should be broken out gently, just above the topmost pair of leaves

in boxes of soil and kept almost dry through the winter in a frostproof place, they will provide still more in spring. To obtain these spring ones the old plants should be cut hard back in February and given more warmth and water together with as much light as possible. The same treatment should be applied to the old plants that are to be repotted and grown on for flowering in the greenhouse.

Geranium cuttings taken in autumn or spring will need one or two stoppings to ensure bushy plants, but no stopping should be carried out after the end of March or the plants will be rather late in flowering. Similar but more frequent stoppings are needed also for the trailing ivy-leaved geranium *Pelargonium peltatum*, cuttings of which are best taken in the late summer and autumn, using short-jointed unflowered shoots about 3in long. Like the zonals, these should then be potted up and grown on until they can be used in baskets, window boxes or other containers in early June.

Penstemons Cuttings of these colourful but comparatively little-known bedding plants are easily taken in August, when unflowered sideshoots 2–3in long will root readily if inserted in sandy soil in a cold frame. Treatment then is the same as for the bedding calceolarias.

Violas The various cultivars of the bedding violas are not used so much these days. They have been replaced to a great extent by the larger-flowered pansies, which are usually grown from seed. Both plants may be propagated by cuttings taken in September and October after the plants have been cut hard back a few weeks earlier. This cutting back will result in the production of basal shoots, which should be taken off about 2in long and treated in the same way as calceolarias.

Chapter 14

Climbers and Trailers

For the most decorative effect in a greenhouse or conservatory some climbing and trailing plants are essential. There is an enormous range in this type of plant, from cool-house ones to those requiring tropical conditions and even some which require no artificial heat at all.

Planting and training climbers

Apart from their decorative effect, climbing plants are useful for providing summer shade. There should not be too many of them in this respect, however, or there will be no suitable place for those plants which need full sun. In lean-to greenhouses, for instance, they should be limited to the back wall and perhaps the ends, while in span-roof structures they are best planted in corners, leaving the centre of the greenhouse relatively open.

To reduce the labour of watering and feeding, the plants should be grown in the actual soil of the greenhouse provided this is drained adequately; or if the structure has a solid floor, it should still be possible to plant into a hole about 18in square broken through this and filled with suitable compost over good drainage. In either case firm planting is essential.

If planting in the ground is impracticable, well-drained tubs or large pots may be used instead, and for some plants of very rampant growth, such as the passion-flower *Passiflora caerulea*, this often proves the better method as it tends to restrict the growth and thus encourage flowering.

Whichever method is used some form of support will be needed, although there are one or two plants, such as *Hoya carnosa*, which have aerial roots and are self-supporting in much the same way as ivy. For the others, wooden trellis or panels of plastic-covered heavy-gauge wire can be used on the back wall of a lean-to structure. If the plants are to be trained along the greenhouse roof they can be tied to wires, preferably galvanised (16-gauge), and run along the roof on the special vine eyes that are obtainable from most good garden shops and which keep the plants away from the glass.

For small pot-grown climbers such as the annual *Thunbergia alata* (Black-Eyed Susan), adequate support can usually be obtained by merely inserting three or four canes in the pot and splaying them outwards. This method can be followed also with some of the larger climbers: the growths are trained round the canes, using wire or twine as an extra support if necessary. Another very effective method is to train the plants like weeping roses over a wire 'umbrella' or a sphere or cone made of wire.

Small plants, including many of the house plant climbers such as *Scindapsus*, *Philodendron scandens* and *Hederas* (the ivies), can be grown over pieces of bark or even logs or branches, to which they will cling with their aerial roots. Another method is to grow them over a column of wire netting filled with sphagnum moss.

Trailers

Plants with a natural trailing or hanging habit can be used to hide the front of the staging and they may also be quite effective if grown on small shelves or corner brackets, where they will have room to hang down without interfering with any plants beneath them. Some are useful too in hanging baskets or in the tubs or large pots in which, say, climbing plants are growing: if they are grown in the actual soil in the tub or pot they will effectively camouflage the container.

In general it is advisable not to grow this type of plant in the ordinary way on the staging, where there is not enough room for the trailing shoots to develop properly. The problem can be overcome to a large extent by standing the plants on inverted pots.

There are a few plants, such as *Helxine soleirolii* (Mind Your Own Business) and one or two of the selaginellas, which can be grown in a shallow bed of soil on the staging, or even beneath it, where they will soon spread and create a more natural effect.

CLIMBERS FOR THE HEATED GREENHOUSE

BOUGAINVILLEA

This showy Brazilian climber, with decorative flower-like bracts, is available mainly as cultivars of two species, *B. glabra* and *B. spectabilis*, with colours in varying shades of pink, red, orange and purple. These make magnificent plants for cultivation either in pots or in the ground: in either case use a mixture of good soil with some leaf-mould and sharp sand. Perfect drainage and full sun are essential. Plant in spring and water freely in summer but keep the plants dry in winter, when they should be pruned by cutting all sideshoots back to two buds as soon as the leaves fall. Winter temperature 45–50° F (7–10° C). Propagate in summer by half-ripe cuttings inserted in sandy soil with bottom heat.

CISSUS

The Australian Kangaroo Vine *Cissus antarctica*, with deeply toothed, dark green leaves about 4in long, is commonly grown as a house plant, climbing by tendrils. If kept safe from frost it is quite easy to grow in any reasonable situation, but the more colourful *C. discolor* is less obliging: it needs much warmth and humidity if it is to retain its large, brilliantly variegated leaves in winter. The J.I.P2 or a soilless compost will suit both plants and propagation may be effected by shoot cuttings, leaf-bud cuttings and layers.

CLERODENDRUM

The Bleeding Heart Vine of tropical West Africa, *Clerodendrum splendens*, grows to about 10ft, with scarlet flowers in summer. The most readily available species, it should be grown in a rich soil with plenty of humus and sand over good drainage. Keep it wet in summer and almost bone dry in winter, when a temperature of 55–60° F (13–16° C) is needed. Shorten the flowered shoots back to 2in or 3in as soon as they are over. Propagate by ripe cuttings in warm close conditions.

HEDERA

Many of the modern cultivars of this plant, the ivy, have become popular in recent years, although more as house plants than greenhouse ones. They are easy to grow as either climbers or trailers in the J.I.P2 and need liberal watering in summer, less in winter. Good light but not full sun is advisable for the variegated kinds, while the green ones will do well in either sun or shade. The indoor kinds may also be grown outside in window boxes etc during the winter if they have first been hardened outside throughout the summer. Propagation is by stem or leaf-bud cuttings, also by layers.

HOYA

The most popular species is *Hoya carnosa*, the Wax Flower, with clusters of pinkish-white flowers, fragrant in the evening, in late summer. It can be grown like ivy against an interior greenhouse wall but also makes a good subject for pots of rough peaty soil. A fairly humid atmosphere is advisable. Liberal watering is needed in summer, but in winter it should be kept almost dry in a temperature of about 45° F (7° C). After the plant has flowered do not remove the flower-stalks as these will produce further flowers the following year. Propagation is by layers or leaf-bud cuttings.

JASMINUM

The fragrant jasmines make excellent plants for the slightly heated greenhouse, two of the best being *Jasminum polyanthum*, a twiner with clusters of pink-tinged white flowers in February, and *J. primulinum*, with fairly large yellow flowers, often double, in early spring. Both are very nearly hardy and do well in any good soil, preferably in tubs or large pots to restrict their growth. Propagation is by cuttings or layers.

MONSTERA

Swiss Cheese Plant *Monstera pertusa*, commonly sold as *M. deliciosa*, is a popular house plant with large perforated leaves and aerial roots by which it clings to its supports. Any good well-drained soil suits it and it should be given ample warmth, about 60° F (16° C) at all times, with liberal watering in summer and less in winter. Pieces of stem will soon root in sandy soil in a close frame at about 75° F (24° C).

PASSIFLORA

The best known species of passion-flower is *Passiflora caerulea*, with flowers consisting of white or pinkish petals and sepals against which is set the corona of filaments, blue, white and purple. This is hardy enough to grow outside in mild areas or in a cold greenhouse elsewhere. Two good species which need rather warmer conditions are *P. antioquiensis*, with rose-red flowers 4in across, and *P. mixta quitensis* (*Tacsonia mixta quitensis*) with large pink to red flowers from July to September. All species will succeed in any well-drained soil in full sun, preferably in tubs or large pots. Prune by shortening back the main stems in early spring, when the resultant new shoots can be taken off about 6in long and rooted as cuttings in close conditions.

PHILODENDRON

Several species of this foliage climber are grown as house plants and they need the same treatment as the Monstera, to which they are closely related.

PLUMBAGO CAPENSIS

Cape Leadwort is an easily grown and beautiful plant, more sprawling than climbing, with clusters of pale blue, phlox-like flowers in summer. No special treatment is needed and the plant can be grown in pots or in the ground: in either case cut back the flowered shoots as soon as they are over. Propagation is heeled cuttings in a warm frame in spring; or pieces of root 1in long can be inserted in a peat-sand mixture in similarly warm conditions.

RHOICISSUS RHOMBOIDEA

Grape Ivy is a fairly common house plant. It is somewhat similar to *Cissus antarctica*, to which it is related, except that the dark green leaves are composed of three leaflets, each rhomboidal as the specific name implies. It needs the same treatment as the Cissus, although if anything it prefers rather warmer conditions.

STEPHANOTIS FLORIBUNDA

Clustered Wax Flower or Madagascar Chaplet Flower is one of the most beautiful of evergreen white-flowered climbers for a fairly warm greenhouse. It will produce in summer a generous supply of very fragrant white flowers, often used in wedding bouquets. Plants grown in the ground are apt to become too big eventually and in small greenhouses they are better grown in large pots, in which they must be kept well watered and fed in summer. A good well-drained soil with some peat or leaf-mould added is all that is needed. In winter a temperature of at least 55° F (13° C) must be maintained and the plants must be kept almost dry. Growth, often slow at first, becomes very vigorous eventually so that constant training in of the shoots is needed to prevent their getting into a hopeless tangle. Propagation is by cuttings inserted in a peat-sand mixture in very warm, close conditions; or established plants will often produce seed, which may be sown in brisk heat in spring.

STREPTOSOLEN JAMESONII

An easy but little known climber with orange trumpet-shaped flowers in profusion in summer. It needs only a frostproof greenhouse and does well in any average soil, with shade from full sun. Plenty of water is needed in summer, but the amount should be reduced towards the autumn so that the shoots can ripen properly before less water still is given in winter. Half-ripe cuttings root readily in gentle warmth.

CLIMBERS FOR THE COLD GREENHOUSE

Most of these plants are on the borderline of hardiness, so that in mild areas they will often succeed outside. Elsewhere, however, they need greenhouse protection, preferably with a little temporary heat during very hard frost.

CAMPSIS RADICANS (syn *Tecoma radicans*; *Bignonia radicans*)

A showy plant, climbing like ivy by means of aerial roots and with scarlet and orange trumpet-shaped flowers in late summer. A good well-drained soil is needed and the plants do better in the ground than in pots. Planting should be carried out in spring and the plant cut hard back annually about the same time. They can be increased by layering, by root cuttings taken in autumn and by cuttings of young shoots in April.

CLEMATIS CALYCINA (syn *balearica*)

Fern-leaved Clematis is an interesting evergreen, bronzy in winter, which produces its creamy yellow, red-spotted flowers from November to February. Like all clematis it does best when its roots are in the shade and it can therefore be planted beneath the staging and strained up through it. Liberal watering and feeding are needed in summer and propagation is by cuttings or layers.

CLIANTHUS PUNICEUS

Parrot's Bill, an evergreen climber with bright red, pea-like flowers, may be grown either in the ground or in large pots, in rich soil over good drainage. Full sun is essential but in hot weather the plants must be syringed frequently and given plenty of water. Cut back the flowered shoots as soon as they are over. Propagation is by seed or by half-ripe cuttings inserted in a sandy soil in a close frame. As the plants dislike root disturbance, both seedlings and cuttings should be potted straight into 5in pots, in which they should remain until they are finally potted or planted.

LAPAGERIA ROSEA

Chilean Bell Flower is one of the finest of all evergreen climbers, with large, waxen, crimson-red bells in summer. It should have partial shade and a peaty lime-free soil with perfect drainage; and it is best grown in a specially prepared bed enclosed by bricks or slates to prevent the spread of the rampant roots. Liberal watering, feeding and syringing are needed in summer, when the plant should be kept as cool as possible. Propagation is by layering or seed.

MANDEVILLA SUAVEOLENS

Chilean Jasmine is a beautiful deciduous climber, with pure white, funnel-shaped flowers that are extremely fragrant. It does better in the ground than in pots, and a mixture of soil, peat and sand suits it. During the summer it should be watered freely, but in winter it needs no water. Pruning consists of shortening the flowered shoots back to about 2in from their base and propagation is by unflowered sideshoots taken about 3in long in summer and inserted in a propagating frame. This plant is perhaps slightly more tender than the other cold greenhouse climbers, and if possible it should be protected from really hard frost.

PASSIFLORA CAERULEA

See list for heated greenhouse, pp 128–31.

SOLANUM JASMINOIDES

Potato Vine is a semi-evergreen with slate-blue flowers resembling those of the potato, to which it is related. These are produced from July until frost puts an end to them. The plant grows easily in any good soil and can be readily propagated by cuttings in summer.

TRAILING PLANTS FOR POTS AND BASKETS

ACHIMENES

See list in Chapter 16 on plants from bulbs, pp 155– 68.

ASPARAGUS

Two relatives of the culinary asparagus make excellent trailing foliage plants for the frostproof greenhouse. One is *Asparagus sprengerii*, with slender arching stems covered with needle-like 'leaves' (strictly speaking, they are not leaves but phylloclades or flattened branches); and the other is *A. plumosus*, a more graceful and plume-like plant whose foliage is commonly used in floral decorations. Both grow easily in any good well-drained soil, with plenty of water in summer and much less in winter. Shade is also advisable in summer. Propagation is by seed or division.

BEGONIA PENDULA

This showy begonia, with masses of small flowers in summer, may be grown from seed, available in mixed colours only, or from tubers, obtainable as named varieties in different shades of yellow, pink, orange, crimson and scarlet. The seed is sown in a temperature of about 70° F (21° C) in early spring. Tubers should be started at the same time by pressing them gently into a mixture of equal parts peat, sand and soil, one tuber to a small pot. Keep them just moist and quite warm until growth is about 2in high, when they can be transferred either to rather larger pots or straight into the baskets in which they are to grow.

CAMPANULA ISOPHYLLA

Star of Bethlehem is an old favourite, with blue starry flowers in late summer. It grows easily in any average soil as long as it is reasonably safe from frost, and it does well either in pots or small baskets. Cut back after flowering and propagate by cuttings or division in spring. There is also a white form, *C. i.* 'Alba'.

CEROPEGIA WOODII

See list in Chapter 16 on plants from bulbs, pp 155–68

COLUMNEA

Two or three species and several cultivars of this plant from tropical America are becoming increasingly popular, but they are by no means easy to grow and some of them are more erect than trailing. They need a temperature of at least 55° F (13° C) in winter, and plenty of light and humidity, if they are to produce their tubular flowers, mostly yellow and red, which are set against small evergreen foliage. A soilless compost suits them and propagation is by cuttings 2in long inserted in a propagating frame at a temperature of about 70° F (21° C).

EPISCIA (Trailing African Violet)

Rare but obtainable, the episcias include some of the most beautiful of trailing plants, but like the columneas they are not easy to grow and require considerable humidity and warmth—at least 60° F (16° C). Their treatment is much the same as that of columneas, except that they need more shade in summer. A soilless compost suits them or they will do equally well in one consisting of soil, peat and leaf-mould in equal parts with a liberal sprinkling of charcoal, crocks and coarse sand. Propagation is by cuttings in a warm frame.

A few of the species offered are *Episcia cupreata*, scarlet; *E. dianthiflora*, a deeply fringed white, and *E. lilacina cuprea*, lavender-blue. All these, like the rest of the family, have superbly beautiful multicoloured foliage.

FUCHSIA

See list of popular plants in Chapter 19, pp 184–95

GAZANIA

See list in Chapter 13 on bedding plants, pp 115–25

HEDERA

See list of climbers above

HELXINE SOLEIROLII

Mind Your Own Business or Baby's Tears is a small-leaved foliage plant which is quite hardy enough to grow in the cold greenhouse. It is easily grown in almost any soil and is readily propagated by division in spring.

HOYA BELLA

This relative of the climbing *Hoya carnosa* is a more shrubby species forming a drooping mass some 18in high, with small evergreen leaves and clusters of waxen white flowers, crimson-purple at the centre. It needs much the same treatment as *H. carnosa* (see list of climbers for the heated greenhouse earlier in this chapter) but requires more warmth in winter, when a temperature of at least 50° F (10° C) is advisable. Propagation is by cuttings or layers, or better still by grafting on to a rootstock of *H. carnosa.*

LOBELIA PENDULA

This is a 'cascade' form of the lobelia used for bedding and it is raised in just the same way. It may be grown also from cuttings taken from plants cut back and kept through the winter. Sapphire and Hamburgia are two good forms.

NEPETA HEDERACEA (syn *Glecoma hederacea*)

An invasive and hardy plant whose use is limited largely to hanging baskets, in which its very long slender stems and kidney-shaped toothed leaves can be quite decorative. It is increased easily by cuttings or division.

PELARGONIUM PELTATUM

See list in Chapter 13 on summer bedding plants, pp 115–25.

SAXIFRAGA STOLONIFERA

Known as Mother of Thousands, Roving Sailor and Strawberry Geranium, this is an old and very easily grown favourite with long runners something like those of the strawberry, and rounded, rather marbled leaves. Its white flowers are small and insignificant. Propagation is by runners.

SEDUM LINEARE VARIEGATUM

An easy-to-grow evergreen plant with 1in long narrow leaves of pale green margined with creamy white. Well-grown established plants are quite effective and are increased very easily by cuttings or division.

SEDUM SIEBOLDII

This tender member of the stonecrop family is grown usually in its variegated form, with fleshy rounded leaves about 1in across and prominently blotched with creamy yellow at the centre. These are produced on pendent stems, about 12in long, which in late summer terminate in clusters of mauve-pink flowers; the stems then die down to leave only the woody root stock for the winter. The plant is easily grown in almost any soil and situation and needs only to be kept safe from frost. Propagation is by cuttings or division.

SELAGINELLA

This enormous family, related to the ferns but more like mosses, provides only a few species hardy enough to be grown in even a cool greenhouse. Two of these are *Selaginella kraussiana*, which forms a mat of bright green tiny leaves, and *S. martensii*, with hard scale-like leaves. Both need warmth and humidity and generally do well in a mixture of light soil with a little leaf-mould and charcoal. Propagation is by division.

TRADESCANTIA (Wandering Jew)

There are several forms of this popular evergreen trailer with narrow pointed leaves often variegated with purple. All are perfectly easy to grow as long as they are safe from frost, but the variegated ones need rather more light than the others. Propagation is by cuttings.

THUNBERGIA ALATA (Black-Eyed Susan)

Unlike the other plants described above, this is an annual, raised easily from seed sown in heat in spring. It may be grown either as a trailer or as a climber; in the latter case thin canes should be used to support the slender twining stems. The fairly large flowers are generally yellow with a purple blotch at the centre, although they can vary widely in colour. They are produced over a long period in summer.

ZEBRINA

This is often confused with Tradescantia, both plants being much alike in appearance and needing the same treatment.

HANGING BASKETS

Baskets for the greenhouse may be made up at any time early in the year as the plants become available, but those for outside are best made up in early May, so that the plants are well established by the time they go out in early June. The traditional method is to line the basket with a thick layer of sphagnum moss, but a more modern method is to use green polythene sheeting with holes punched in it for drainage. A saucer placed in the bottom of the lining helps to reduce the amount of watering needed.

The basket is then partly filled with a good compost and the roots of a few trailing plants such as *Lobelia pendula* are pushed through from the outside and spread over this before being anchored firmly with a further layer of compost. The main plants, generally from pots, are then placed in position with one or two erect ones at the centre and the trailing ones towards the edge, and packed round with compost to bring the surface to about 2in from the rim. A few nasturtium seeds pushed into the compost will help eventually to enhance the display.

For small baskets a good display may be obtained by putting in only one type of plant, and *Begonia pendula*, the trailing forms of fuchsia and achimenes are particularly effective when used in this way. For both small and large baskets a stout hook is needed and watering must be attended to continually.

Chapter 15

Foliage Plants

Most people no doubt think of greenhouse plants in terms of flowers but there are many which, although not producing much in the way of bloom, make up for this by their decorative and often colourful foliage, far more durable than flowers. Another of their advantages is that nearly all of them are perennial and can be carried on from year to year. When they do eventually become too large or too old, it is usually fairly easy to renew them by cuttings or by other means of vegetative propagation, or in some cases by seed.

Of the two main groups of foliage plants—the green ones and the multicoloured—the former are generally the hardier, in most cases requiring only enough heat to keep out frost. The coloured ones, on the other hand, are largely tropical in origin and thus need considerably more warmth, say a minimum of 50° F (10° C) in winter; some need an even higher temperature than this. This group tends to need more humidity too, but given this they are not much more difficult to grow than the green ones, which by and large are among the easiest of all greenhouse and indoor plants.

Ferns

Although ferns obviously come into the class of foliage plants, they constitute a distinct group in themselves. During the summer they need copious watering and a shady position. Apart from a few species with hairy or powdered leaves, they also appreciate frequent syringings both overhead and around the pot, carried out preferably in the evenings of warm days. Then from early autumn onwards no further syringing should be done. As the plants are resting at that time they can be kept fairly dry, but not to the extent of being allowed to dry out completely. This is particularly important with the evergreen species: if these are kept too dry, they will lose their leaves.

The repotting of mature established ferns is needed only when the pots are absolutely full of roots, for ferns do best when they are practically pot-bound and the inside of the pot is covered with a network of fine roots. Young plants, however, are best repotted fairly frequently so that their growth is not checked by starvation, but this should be done in stages, moving them on, say, from 3in to 4½in pots, then to 6in ones and finally to 8in.

Repotting is best carried out in the spring, when the plants are just restarting into growth after the winter. Plastic or clay pots may be used, although for some reason or other ferns seem to do best in old clay ones; in either case it is important to make sure that the pots are well crocked to at least a quarter of their depth. The plants, which should be quite moist at the root, should have their matted roots loosened slightly before being potted moderately firmly into either the J.I.P1 or 2 according to their age and size. A simpler compost that will suit most popular kinds is 2 parts soil and 1 part each of peat and leaf-mould, to which has been added a good sprinkling of charcoal, coarse sand and very small crocks. After potting, water in the plants and stand them in the shade, where they should be kept barely moist until growth is seen to have started again. But do not let them get too dry as, unlike most plants, ferns seldom recover fully if they are allowed to flag.

Propagation

Most of the popular ferns are easily propagated by division in spring. They should be separated into fairly large portions, as these tend to be more successful than smaller ones. Some kinds with spreading rhizomes (horizontally spreading stems that produce roots and shoots) may be propagated also by layering and in this case the portions to be dealt with may be pegged down into small pots of suitable compost placed next to the old plant.

Page 141 (*right*) The beauty of some succulents, such as *Pachyphytum oviferum* shown here, lies in the attractive bluish-white bloom that covers the leaves;
(*below*) an easy and popular succulent, *Aloe aristata*

Page 142 (*left*) The rich blue *Scilla sibirica* Spring Beauty is grown easily from bulbs to flower in the greenhouse or indoors in early spring; (*below*) *Pteris cretica* is an easy-to-grow fern, sometimes included in bowls of bulbs at Christmas

Propagation by spores

Being non-flowering plants, ferns do not produce seeds but they do produce spores which can be sown in much the same way. It is possible to buy these—some are usually listed in good seed catalogues—but where mature plants are already being grown it is not difficult to save them oneself.

On most types of fern the spores are produced on the back of the fronds, in minute capsules known as sporangia which in turn are grouped into clusters called sori. These sori can usually be seen easily as a definite pattern of dots or lines, and when they darken in colour this is an indication that the spores are ready for dispersal.

It is at this stage that they can be collected and a simple method of doing this is to lay a mature frond face upwards (that is, with the sori downwards) on a sheet of glossy paper, with a piece of blotting paper weighted down lightly over it. After a few days the spores will have been discharged on to the lower paper, from which they can be poured carefully into a small bottle or other container.

Sowing the spores

This is no more difficult than sowing tiny seeds. A perfectly clean seed pan should be well crocked and filled moderately firmly with a mixture of moist peat and sterilised soil, topped with a thin layer of finely sieved peat. The spores are then sown thinly on the surface and the pan is stood in a saucer of water placed in a warm shady position and kept filled. To create the necessary humidity, the pan is covered with a piece of glass, which must be turned daily to get rid of the condensation.

Prothalli

Unlike seeds, spores do not produce young ferns immediately. The first sign of growth is no more than a film of green over the surface, consisting of tiny plants known as prothalli. With most of the popular ferns, these prothalli appear within a few weeks of sowing, and as soon as they can be identified as tiny plants, they must be pricked out in small clusters no more than ¼in across, into the same sort of compost used for the sowing of the spores. This is a delicate job, each cluster being pressed lightly into the compost. The pricked-out plants are again stood in a saucer of water and covered with glass until the tiny ferns can be seen. The glass can then be removed and the plants grown on until they are large enough to be potted up into 2in or 2½in pots.

LIST OF FOLIAGE PLANTS

ARALIA ELEGANTISSIMA

See DIZYGOTHECA

ARALIA SIEBOLDII

See FATSIA JAPONICA

ASPARAGUS

See list of trailers in Chapter 14, pp 126–38

BEGONIA REX

The many excellent cultivars of this are among the finest of foliage plants, with large multicoloured leaves. Warm humid conditions, with a minimum of 50° F (10° C) in winter, are needed and the plants do best in the shade, even beneath the greenhouse staging so long as there is no drip. Propagation is by leaf-cuttings.

CALADIUM

This tuberous plant from tropical America is available in several different forms, all making magnificent plants about 1ft high with large, very colourful and mostly arrow-shaped leaves. Small plants make good table decorations for temporary use. Start the tubers in March by planting them in small pots of moss and keeping them well syringed; then when growth is well started pot them on into 5in pots, using a compost consisting largely of humus over perfect drainage. Copious watering and very warm humid conditions, with shade only from the hottest sun, are needed in summer. In autumn the plants should be dried off gradually and rested through the winter in a warm place, where they should be given just enough water to prevent them from shrivelling and rotting. Propagation is by division of the old plants or by seed sown in brisk heat, although this seems to be rarely available.

CALATHEA

See MARANTA

CHLOROPHYTUM ELATUM VARIEGATUM

Spider Plant is one of the easiest and most popular of house plants, with long arching leaves striped green and cream. Almost any soil seems to suit it and as long as it is safe from frost it demands no more than liberal watering and feeding in summer, with rather less water in winter. Browning of the leaf-tips is usually caused by a too dry atmosphere. Propagation is by division or by means of the small plantlets that develop on the ends of the flower-stems.

CODIAEUM (Croton)

One of the finest foliage plants for the greenhouse with a minimum winter temperature of 55° F (13° C). *Codiaeum variegatum* is the only species grown, but this is available in many fine forms with leaves which are splashed and mottled with various colours and which vary in shape from broad to strap-like. The plants need plenty of watering and syringing in summer, when it is almost impossible to keep them too warm; less watering is needed in winter, although they should never become dry. Small plants suitable for temporary indoor decoration can be raised by cutting the tops off existing plants and rooting these in small pots of the J.I.P1 in a close frame, with a temperature of at least 70° F (21° C). Unlike most cuttings, they should be watered fairly freely while they are rooting. Propagation can be effected also by leaf-bud cuttings, in similarly warm conditions.

COLEUS

In its many variously coloured forms *Coleus blumei* is perhaps the most popular of all the variegated foliage plants for the greenhouse. Seeds sown in spring in a temperature of 65° F (18° C) will produce good plants the same summer if potted on into 3in and then 5in pots of the J.I.P1. Stop the plants at four pairs of leaves and the resultant sideshoots at two pairs. A warm moist atmosphere is needed and the plants should be shaded from full sun, which would bleach the leaves.

The most colourful plants from these seedlings may be perpetuated by cuttings taken 2–3in long in August and inserted in sandy soil in a warm frame. These must then be wintered fairly dry in a warm greenhouse.

CORDYLINE

See DRACAENA

CROTON

See CODIAEUM

CYPERUS ALTERNIFOLIUS (Umbrella Plant)

This is an easy plant to grow and deserves to be better known than it is, with erect green stems 2ft or more high terminating in grass-like leaves which arch out from the centre. It may be grown from seed sown in heat in spring or propagated by division; in either case grow the plants on in warm conditions with liberal watering. Established plants are best stood permanently in water.

DIEFFENBACHIA

Different forms of *Dieffenbachia picta*, all with green leaves variously marked with cream or vory, are available but, owing to their poisonous qualities and habit of shedding their lower leaves, they are not ideal plants for the home or greenhouse. Their treatment is the same as for codiaeums.

DIZYGOTHECA ELEGANTISSIMA (False Aralia)

This greenhouse shrub has become increasingly popular as a house plant in recent years, but it is not one of the easiest to manage, as it needs a temperature of at least 60° F (16° C) in winter and also a draught-free humid atmosphere. When well grown it makes a plant 2–3ft high, with brownish-red young leaflets—rather like long and very narrow oak-leaves—which radiate like the spokes of a wheel from the tips of the erect mottled stems. As the leaflets age, they turn a rich dark green. The plant thrives in any well-drained compost with plenty of humus. It should be watered liberally in summer and rather less so in winter. Propagation is by cuttings in a warm frame in summer or by root cuttings in similar conditions in spring.

DRACAENA

The two genera *Dracaena* and *Cordyline* are often confused and the most popular species, usually sold as *Dracaena terminalis*, is in fact *Cordyline terminalis*. Several named forms of this colourful evergreen plant are available, all making superb plants which become somewhat palm-like as they age. The leaves, up to 12in long, are splashed or otherwise marked with brilliant colours, although unfortunately the full beauty of the colouring does not develop until the plants are fairly old. They need high temperatures and liberal watering in summer, while in winter they should be kept rather dry in a temperature of at least 55° F (13° C). Fairly large pots will be needed as the plants grow and the J.I.P2 is a suitable compost to use when repotting.

The plants may be grown from seed sown in a temperature of 85° F (30° C), or established plants may be increased by cutting the stem into portions and inserting these in a peat and sand mixture in a very warm frame. Some cultivars may also be increased by buds, known as 'toes', which are produced on the underground stems. These should be inserted in the same way as stem cuttings.

FATSIA JAPONICA

This evergreen, with its large, glossy, palmate leaves of dark green, is easy to grow. It is hardy enough to be grown in a cold greenhouse or in a cold room or hall indoors; then when it eventually becomes too large it may be safely planted out in the garden in a sheltered spot. It is best propagated by seed, sown in heat in early spring, which will produce good house plants by the end of the year.

FICUS

Several species in this Fig family have become popular as house plants, notably the Indiarubber Plant *Ficus elastica*. Frequent sponging of the large leaves and liberal watering will help to keep this plant healthy in summer, but in winter, when it will be quite safe in a temperature of about 50° F (10° C) it must be allowed to dry out between thorough soakings. In time the lower leaves may fall, and it is advisable to raise a new plant from a cutting or layer as described below. The original plant should then branch out and develop a bushy shape.

FICUS BENJAMINA

This is less popular, although it makes a most attractive little tree-like plant with graceful pendent branches covered with pale green leaves which darken in colour as they age. It needs plenty of water in summer and rather less in winter, when it should be kept in a temperature no lower than 50° F (10° C). The creeping *Ficus pumila* (*repens*), with small green leaves, also needs a good deal of water; in fact it is fatal to allow it to dry out. Apart from this it is quite easy to grow, its only other requirement being that of almost complete shade. *F. radicans*, another creeper, with larger leaves than *F. pumila*, is most commonly grown in its variegated form with green and cream foliage. This again is a plant that must never be allowed to dry out. It needs very much warmer and more humid conditions than *F. pumila* which makes it more difficult to keep in good condition through the winter.

The propagation of the *Ficus* species differs according to the type of plant. *F. elastica*, for instance, may be propagated by air-layering, which means simply that the top of the stem is rooted while it is still on the plant. The uppermost leaves are first tied together to keep them out of the way and then after removing those immediately beneath them a diagonal cut, sloping upwards and inwards for about 1½in, is made into the bared portion of stem. This cut portion is then packed round with sphagnum moss covered in turn with polythene sheeting, tied at the top and bottom round the stem. Eventually roots will appear in the moss and as soon as these can be seen the rooted top portion of the stem may be cut off and potted up.

F. elastica may also be propagated by leaf-bud cuttings and by terminal shoot cuttings. Both types should be inserted in small pots containing a peat and sand mixture, the pots then being placed in a warm frame at a temperature of 70–75° F (21–24° C).

F. benjamina is best increased by heeled cuttings inserted in a pot of peat and sand in similarly warm, close conditions. The trailing species may be increased by leaf-bud or shoot cuttings; alternatively, the trailing stems, which produce aerial roots, may be layered into small pots of compost.

FITTONIA

Two species of this plant are suitable for cultivation indoors or in a warm greenhouse. One is *Fittonia argyroneura*, with rounded leaves net-veined in white, and the other is *F. verschaffeltii*, with more oval leaves netted with carmine. Both are creeping plants which need an open compost with plenty of humus and do best if moist in warm shade. Never let them dry out. Propagation is by short terminal cuttings in a warm close frame. Young plants should be stopped to encourage a branching habit.

GREVILLEA ROBUSTA

This Australian Silky Oak makes an attractive little palm-like plant with finely cut grey-green leaves. It is raised easily from seed sown in heat in spring and should be grown in the J.I.P1 in a warm buoyant atmosphere with shade from full sun. It makes a good house plant.

HOWEA

A family of palms, of which two species, *Howea forsteriana* and *H. belmoreana*, are popularly grown under the name of Kentia Palms. They are raised from seed sown in a temperature of at least 80° F (27° C) and should be grown on by being potted on in gradual stages, using the J.I.P1 and eventually the J.I.P2. A warm humid atmosphere with partial shade suits them. They should be watered freely in summer and rather less in winter, when a temperature of about 60° F (16° C) is advisable.

IRESINE

See list in Chapter 13 on summer bedding plants, pp 115–25

MARANTA

The family Marantaceae includes three genera which are usually grouped together. These are *Maranta*, *Calathea* and *Ctenanthe* and they all include a number of attractive foliage house plants. Being of tropical origin they need ample warmth, humidity and shade, together with a fair amount of water at all times. Propagation is by division or cuttings.

PEPEROMIA

Several species of this plant are grown as house plants for the sake of their decorative leaves, which are more or less fleshy and attractively shaped and coloured. The plants also produce whitish flowers on a spike (spadix) resembling a pipe-cleaner, but these are more a curiosity than anything else. A temperature of 45–50° F (7–10° C) suits most peperomias and they should be grown in a compost consisting mainly of humus, with only enough water to prevent them from drying out completely. No water at all should be given when the plants are growing in cool conditions in winter. Propagation is by leaf or stem cuttings according to the species.

PHILODENDRON

See list of climbers in Chapter 14, pp 126–38

PHOENIX

A family of palms, some of which may be raised from seed to provide very decorative pot plants. Treatment is as for *Howea*. *Phoenix dactylifera*, the Date Palm, may be easily raised from the stones, kept moist and warm.

RHOICISSUS

See list of climbers in Chapter 14, pp 126–38

SANSEVERIA

Sanseveria trifasciata laurentii, with stiff erect leaves banded and margined with yellow, is the house plant commonly known as Mother-in-Law's Tongue or Bowstring Hemp. It is easily grown in moderate warmth in sun or shade provided it is allowed to dry out between waterings, particularly in winter, when one soaking will probably be enough for a month or more. It can be propagated by leaf cuttings, but in this case much of the variegation will be lost: to preserve this, the plant must be raised from suckers ('toes') potted up and kept in warm close conditions.

STROBILANTHES DYERIANUS

A little-known plant, but one which is very beautiful in its young state although it becomes straggly with age. The long, oval, slender-pointed leaves have almost the appearance of a bird's wing: the basic dark green colour of the upperside is overlaid with a metallic blue and purple sheen, while the underside is a bright purple. The plant needs warmth, humidity and shade and is readily propagated by cuttings in heat.

TRADESCANTIA

See list of trailers in Chapter 14, pp 126–38

ZEBRINA

As for TRADESCANTIA

SOME POPULAR FERNS

ADIANTUM

This is the large family of Maidenhair Ferns of which only a few are available for cultivation. *Adiantum cuneatum* is the one most commonly grown and this should do well in any open porous compost which is not made too firm, provided it is given partial shade and liberal watering in summer. Even in winter, when a temperature of not less than 50° F (10° C) is advisable, the plant should not be allowed to dry out. Propagation is by division in spring or by the spores, which often sow themselves in the greenhouse.

ASPLENIUM

Two species of this fern are commonly grown. One is the Bird's Nest Fern *Asplenium nidus*, which hardly looks like a fern at all with its erect undivided fronds springing from a central crown. The other is *A. bulbiferum*, a quite graceful plant which produces young plantlets on its much divided fronds. Both make good evergreen plants for the house or greenhouse as long as they are given filtered shade, liberal watering in summer and a well-drained humus soil. Keep the plants just moist in winter, when *A. nidus* needs rather warmer conditions than the other. Propagation is by young plantlets in the case of *A. bulbiferum*, and by spores or division for *A. nidus*.

CYRTOMIUM FALCATUM

This evergreen Japanese Holly Fern makes a good house plant for a cool shady position. Although not one of the most handsome of ferns, it can be quite effective with its tall erect fronds which consist of a number of oval, very pointed divisions attached to the main stalk of the frond, something like the leaflets of a pinnate leaf. It grows readily if kept moist in a humus soil to which has been added some coarse sand or small crocks, and it is propagated easily by division or by spores.

DAVALLIA CANARIENSIS

The Hare's Foot Fern is so called on account of the brown hairy rhizomes which often trail over the edge of the pot. It makes a good plant for a basket or pot if watered liberally in summer, with less water in winter, when it needs only enough heat to keep it safe from frost. Any well-drained compost consisting largely of humus suits it, and propagation is by division or spores.

NEPHROLEPIS

This genus includes many fine greenhouse ferns, the best known being *Nephrolepis exaltata*, the Boston Fern, a beautiful species for a hanging basket, with long arching fronds up to

3ft long. A suitable compost is 3 parts peat to 1 each of soil and sand. The plant should be kept moist in this at all times, but more so in summer. Light shade and a minimum temperature of 50–60° F (10–13° C) are also needed. Propagation is by spores or by division.

PELLAEA ROTUNDIFOLIA

An excellent plant for a small hanging basket, particularly as it needs less water than most other ferns. It does well in a compost of 2 parts peat to 1 each of soil and sand, with a generous sprinkling of small crocks to ensure good drainage. It should be kept in full light, with shade from only the hottest sun, in a temperature no lower than 45° F (7° C). Propagation is by spores or by division.

PLATYCERIUM BIFURCATUM (Syn *P. alcicorne*)

This Stag's Horn Fern is a particularly interesting one to grow as it is quite different from the others. The common name is taken from the branching fertile fronds (those that bear the spores) which radiate from a cluster of rounded barren ones—the 'mantle' or 'shield' fronds—which in nature enable the plant to cling to the tree on which it is growing. (Unlike the other ferns mentioned here this is an epiphyte, which means that it grows on trees but not as a parasite.) These mantle fronds serve also to gather the fallen leaves and other debris in which the roots take hold, which gives some ideas as to the sort of compost the plant needs. Equal parts sphagnum moss and rough peat with a little sand added make a suitable mixture if the plant is being grown in a pot or basket. Another and more effective way of growing it is to wrap the roots in sphagnum moss and peat and then to tie the whole thing to a block of wood. If the plant is then suspended it will grow outwards in all directions and make a most unusual feature of the greenhouse.

The treatment of this fern also differs to some extent from that of the others. The plant does best in filtered shade, with frequent syringings to create a humid atmosphere, but the actual watering should be limited to thorough soakings given only when the fronds start to droop from dryness: this is one of the few ferns which will stand this drastic treatment. A temperature of 50–55° F (10–13° C) is adequate, and propagation is carried out by spores or by division, or by means of the suckers or young plantlets which appear on the roots.

PTERIS

This genus provides some of the most easily grown and most popular of all ferns; in fact, some of them are often included as small plants in the bowls of flowering bulbs sold at Christmas. The most common are the forms of *Pteris cretica*, which yields a profusion of much divided fronds, often crested or curled, 12in or so high. It grows easily in the J.I.P1 and as long as it is kept safe from frost, without drying out, it should present no problem. Propagation is by spores or by division.

Chapter 16

Plants from Bulbs

Many of the most attractive and interesting plants are raised from bulbs, corms, tubers and similar growths, all of which, although botanically different, may here be regarded as 'bulbs'. Some of these, such as the hyacinth, daffodil and tulip, are well enough known but there are many others which between them can provide a show for almost the whole year. There is also the advantage that most of these bulbous subjects can be kept from year to year, in many cases increasing themselves by offsets or bulblets that can be taken off and grown on as new plants.

Some may also be grown from seed, a particularly useful method with, for instance, the begonia and gloxinia which can be flowered in the same year as they are sown. On the other hand, some of the plants take several years to flower from seed and in this case, obviously, the vegetative method makes a more satisfactory way of propagation.

It is, however, almost impossible to generalise about the cultivation of the bulbous plants. Further details on the individual ones are therefore given in the list below.

ACHIMENES (Hot Water Plant)

Erect or trailing plant, with funnel-shaped flowers in a wide range of colours, for pots or hanging baskets. Start with tiny scaly rhizomes or tubercles into growth in spring in a temperature of 60° F (16° C) and keep them barely moist until growth begins. The compost should be well drained and consist mainly of rich humus, with some coarse sand and soil added. For baskets the tubercles can be started by planting them 1in deep and 2in apart in seed trays, from which they can be transferred to the baskets as soon as growth is about 2in high. For pot culture they can be started in the actual pots in which they are to flower, set at the same depth and spacing as in the boxes. It is said that pouring hot water on the soil helps to start the plants into growth—hence the common name—but this is not necessary if the temperature is adequate.

Water the plants freely in summer and shade them from full sun; then in autumn dry them off gradually and store them, still in the compost, in a temperature not less than 50° F (10° C), with no water at all. Propagation is carried out easily at replanting time in spring, when the tubercles will have increased in number and can be separated.

BEGONIA

The tuberous begonias with large double flowers are among the easiest to grow and the most showy of summer-flowering greenhouse plants. They may be grown from either seed or tubers, but the latter method is by far the simpler.

Seed is sown in the early spring in a temperature of at least 60° F (16° C) to produce flowering plants the same year. It is extremely fine and should be sown merely on the surface of the compost. Then before the seedlings become overcrowded they should be pricked out and potted eventually into 3in pots of the J.I.P1 and finally into 5in or 6in ones of the J.I.P2.

Tubers are started by planting them close together in a box containing a mixture of equal parts sand and moist peat or leaf-mould so that they are just covered. Keep them barely moist in a temperature of 55–60° F (13–16° C) and when the plants have made 2in or so of growth pot them into 5in or 6in pots of the J.I.P2, with good drainage. Water them sparingly to start with. A fairly humid but buoyant atmosphere and shade from full sun will then encourage good growth and as soon as the flower-buds appear, the small female ones on either side of the main central (male) ones should be removed.

The plants may also be used for bedding by starting the tubers off in boxes in March and growing them on until they are planted out in early June in a good well-drained soil with some lime added.

Another method is to plant the dormant tubers straight into the ground at the end of April, setting them 2in deep and 12in apart. The tubers may also be planted in window boxes and other plant containers outside.

Begonia tubers may be propagated by division and by cuttings. For the former, the tubers are started into growth in the normal way but as soon as the 'eyes' (the pink buds which denote new growth) can be seen, the tubers are cut into portions each containing at least one eye. The cut surfaces are then dusted with sulphur and the divisions are started off in the same way as the complete tubers.

Cuttings may be taken in two ways. One is to use the young shoots which appear when the tubers are started into growth: one or more may be severed from each tuber provided two or three are left intact to flower. Those selected for use as cuttings should then be cut out about 3in long with a tiny piece of the tuber and rooted as normal cuttings. For a more rapid increase sideshoots consisting of the unflowered shoots which appear in the axils of the leaves (where the leaf joins the stem) may be used. These are taken when they are about 4in long in summer. Sever them with a V-shaped cut right at their base (see Fig 7) so that the base of the cutting includes the small bud or growth that will be there already, for this is the bud which will produce the main stem the following year. The cuttings should then be rooted in the normal way.

CALADIUM

See list of foliage plants in Chapter 15, pp 139–54

CANNA (Indian Shot)

Canna indica and its hybrids, all about 3ft high with gladiolus-like flowers and large green or purplish leaves, make fine plants for the greenhouse or for bedding, although for the latter purpose they usually do well only in very sunny areas. The plants are grown from fleshy rhizomes which are stored through the winter in much the same way as dahlia tubers, except that they should be set in damp peat to prevent them from drying out completely.

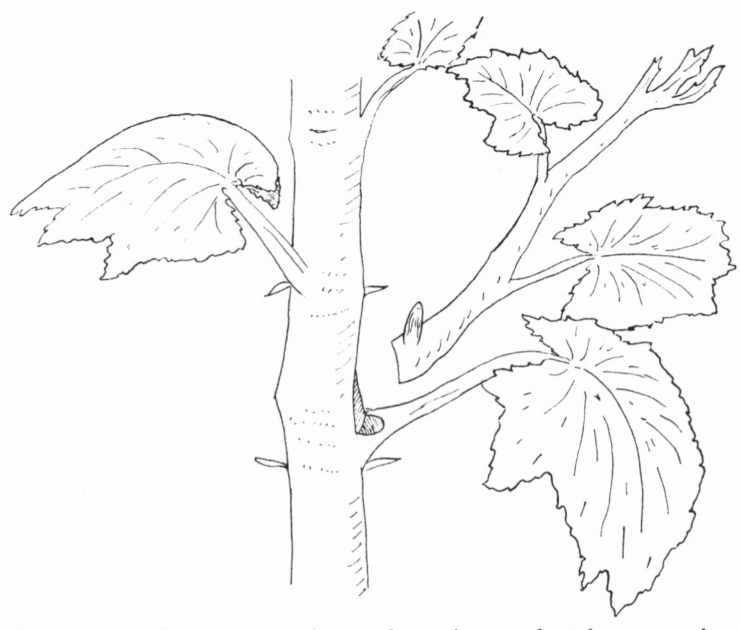

Fig 7 The sideshoots on tuberous begonias may be taken as cuttings by severing them with a V-shaped cut at the base, taking care to include the small dormant bud there

Then in spring they are started into growth with more water and warmth and potted up eventually into 5in or 6in pots of rich compost for greenhouse display.

Alternatively, when the plants have been started into growth, single shoots with a rooted portion of rhizome attached may be taken off and potted up for bedding out in early June.

The plants may be grown also from the large black seed which gives the plant its common name. The seed should be steeped in warm water for forty-eight hours before being sown in a temperature of at least 70° F (21° C).

CEROPEGIA (Rosary Vine)

The only species commonly grown is *Ceropegia woodii*, a trailer with very long slender stems, small, heart-shaped marbled leaves and tiny purplish flowers. It grows from a corm but is usually

Page 159 A fine display of *Primula obconica* in spring

Page 160 (*above*) Schizanthus hybrids in summer; (*below*) a summer greenhouse, with a variety of plants making the most of the available space

bought as a house plant, when propagation can be carried out by means of the minute corms that are produced at the joints of the stems. It thrives in the J.I.P1 with some extra grit added, provided it is not over-watered: this means giving no water at all in the winter and not much more in summer. A temperature of 45° F (7° C) is high enough in winter.

CLIVIA (Kaffir Lily)

The species most commonly grown is the summer-flowering *Clivia miniata*, with evergreen strap-like leaves and large heads of funnel-shaped flowers usually red or yellow, although there are several different forms of this plant. The bulbs seem difficult to obtain but seed is available and this will produce flowering plants after eighteen months or so. Sow it in a temperature of 70° F (21° C) and keep it quite moist; then pot up the seedlings in stages, using a good soil with some peat or leaf-mould and a little charcoal and bonemeal added. Keep warm and wet in full light in summer but almost dry in winter, when a temperature of 45° F (7° C) is adequate. Repotting of established plants is seldom needed and propagation, by division at repotting time, is difficult owing to the tangled nature of the fleshy roots.

CRINUM (Cape Lily)

Crinum powellii is hardy in mild areas but elsewhere needs the protection of a cold or slightly heated greenhouse. There are both pink and white forms, with funnel-shaped flowers on stems 3ft high and large strap-like leaves. Pot the bulbs in spring and treat as for *Clivia*. Propagation is by offsets.

CYCLAMEN

This extremely popular plant is grown initially from seed but the corms can then be kept from year to year. The seed may be sown at any time from August to January, but the finest plants are produced from sowings made early in this period, when it is easier to maintain the temperature of 60° F (16° C) needed for germination. For the later sowings it is advisable to soak the seed for twenty-four hours and then to sow it after it has dried a little.

Either the J.I.P1 or a soilless compost may be used as the sowing medium, the seed being set 1in apart and ¼in deep. If kept quite moist it should germinate in about a month, when the seedlings should be grown on in a temperature of 55° F (13° C) in a position close to the glass. As soon as they have made two or three small leaves, prick them out so that the tiny corms are just covered and keep them moist but not wet, with a little extra warmth, until they start to grow again. Transfer them to 5in pots of the J.I.P1 as soon as the leaves are touching in the pricked-out boxes and put them finally into well-drained 5in pots of the J.I.P2.

The essentials for success with this plant are correct watering, a well-drained porous compost not made too firm, and full light, with shade only from strong sun. During the early stages of growth, particularly when the plants have been pricked out or potted, it is essential to keep the compost only just moist enough to encourage root growth, but when strong growth is being made the plants will take liberal watering. This applies even in winter when they are in bloom, provided the temperature is about 55° F (13° C), but in cooler conditions the best method is to allow them to become almost dry between thorough soakings, applied preferably from underneath.

After they have flowered, keep the plants well watered in full light until the leaves start to turn yellow, then dry them off and keep them bone dry in full sun until about July or early August. The corms can then be shaken out of the old soil and repotted.

DAFFODIL

See NARCISSUS. The treatment for bulbs grown in undrained bowls indoors is much the same as for boxed ones, but bulb-fibre should be used in place of soil. The bowls are best kept in a cool, dark and airy place indoors until growth is an inch or two high, when they can be brought into more light and warmth.

DAHLIA

Although these are actually garden plants, they are usually stored and started into growth in the greenhouse. When the frost has blackened the foliage in the autumn, lift the plants and cut the stems down to a few inches high; then after cleaning and drying the tubers, store them loose in open boxes in a cool dry corner of a frostproof greenhouse.

In spring, place them close together in boxes or on the open greenhouse bench and partly cover them with damp peat so that the upper parts of the tubers are left exposed. In moderate warmth they will soon send up new shoots and these can then be taken off and rooted as cuttings to be potted up for planting out in early June.

FREESIA

See Chapter 12, pp 107–14

GESNERIA CARDINALIS (syn *G. macrantha*)

The Cardinal Flower is a tuberous-rooted plant with velvety green leaves and terminal clusters of scarlet tubular flowers in late summer. It is a relative of the gloxinia and should be grown and propagated in much the same way.

GLORIOSA (Glory or Climbing Lily)

The most commonly grown species of these plants from tropical Africa is *Gloriosa superba*, which climbs to about 6ft by means of leaf-tendrils and produces showy scarlet and orange flowers in summer. The brittle tubers, which must be handled carefully to avoid damage, should be potted in February, each one set 2in deep in a well-drained pot containing a mixture of soil and peat. A temperature of about 70° F (21° C) is needed to start them into growth, when more water should be given gradually, until flowering is over. Then the plants should be allowed to die down before being stored in their pots in a warm place, where they should be kept quite dry.

GLOXINIA

This popular greenhouse plant, with showy trumpet-like flowers in a wide range of colours, is grown both from tubers and seed. The former should be started in early spring by being planted so that they are just covered in a spongy peaty soil, either singly in small pots or close together in a shallow box. Keep them barely moist in a temperature of about 65° F (18° C) until the new shoots are 2in or so high, then pot the plants on into 6in pots of the J.I.P2 or a soilless potting compost. A fairly humid atmosphere and shade from full sun are needed in summer. When flowering is over in autumn, the plants should be dried off gradually and stored quite dry in a temperature of not less than 50° F (10° C).

The seed is sown usually in early spring to provide flowering plants for the same year, and as it is extremely fine it should be sown merely on the surface of the compost. The tiny seedlings should be pricked out as soon as they are large enough to handle and potted eventually into 3in pots in which they should be grown on as above. Propagation may be carried out also by leaf cuttings.

HAEMANTHUS KATHERINAE

A showy African plant, with large globular heads of bloom, something like a huge scarlet dandelion 'clock' up to 7in across. It needs much the same treatment as the hippeastrum (see below), except that the bulb should be buried completely. Propagation is by offsets.

HEDYCHIUM GARDNERIANUM (Ginger or Butterfly Lily)

A good plant, with large spikes of fragrant lemon-yellow flowers on 4ft stems, for the cool greenhouse or for flowering outside in summer. In either case large pots or tubs should be used, with the rhizomatous rootstocks being planted in these in spring, in a rich well-drained soil with plenty of humus. After flowering cut the flowered stems down and keep the plants safe from frost and almost dry throughout the winter. Propagation is by division in spring.

HIPPEASTRUM

Often mistakenly called Amaryllis, this plant in its many hybrid forms has become very popular in recent years as both a greenhouse and an indoor plant. The large trumpet-like flowers, in a wide range of colours, are carried on strong bare stems above strap-like foliage, and although they are usually produced from spring onwards, earlier ones can be obtained by using 'prepared' bulbs potted up in early winter. For normal cultivation the bulbs are potted in spring, each one set very firmly to half its depth in the J.I.P2 in a well-drained pot just large enough to take the bulb comfortably, with about 1in to spare all round. After potting, keep the plants warm and almost dry until growth starts, when more water should be given gradually. After flowering, remove the flower-stems and grow the plants on with plenty of water, food and light for the rest of the summer; in the autumn dry them off and store them without water through the winter, safe from frost. Propagation is by offsets when repotting, which is needed only every three or four years.

HYACINTHUS (Hyacinth)

For greenhouse and indoor flowering, the bulbs should be grown preferably in undrained bowls of bulb-fibre, with a start being made in September for Christmas flowering. Put a layer of the fibre in the bowl, place the bulbs on this with about 1in between them and fill the bowl with more fibre until the 'nose' of each bulb is about 1in above the surface. Do not make the fibre too compact, particularly beneath the bulbs. The plants should then be kept quite moist in a cool, dark and airy place until growth is about 2in high, when they should be brought into the light and given more warmth and water. For flowering at Christmas specially 'prepared' bulbs can be obtained.

Propagation is seldom practised on an amateur scale but forced bulbs which have been planted out in the garden will eventually produce offsets that can be taken off and grown on in a nursery bed to flowering size.

LACHENALIA (Cape Cowslip)

This plant is available in several forms such as *Lachenalia nelsonii* and *L. luteola* (both with yellow flowers marked green), *L. glaucina* (white to deep violet) and *L. pendula* (coral red edged green and purple). All these have tubular flowers carried in drooping spikes and they make fine plants, 6–12in high, for winter and spring if given the same treatment as FREESIAS (see Chapter 12, pp 107–14).

LILIUM (Lily)

Some of the hardy lilies such as *Lilium regale*, *L. henryi*, *L. hansonii*, *L. speciosum* and *L. auratum* make excellent plants for the cold or slightly heated greenhouse. A particularly good show may be obtained by growing several of the same sort together in a large pot or tub. Alternatively, single bulbs may be grown in 6in pots or very large ones in 8in pots. Perfect drainage is essential and a suitable compost is 2 parts rough soil and 1 part of leaf-mould or moist peat, with a little charcoal and enough coarse sand to keep the mixture open.

Lily bulbs are of two kinds: those which produce stem roots and those which do not. The bulbs of the former should be set well down in the pot and just covered with compost to allow for top dressing later, but the latter should be similarly covered with compost at the normal level in the pot. After potting, stand the plants in the shade outside and cover them with 6in of peat to protect them from heavy rain. Little or no watering is needed until growth starts, when the plants should be removed to the greenhouse and given more water gradually. After flowering, keep the plants cool and barely moist in a frame until it is time to repot and start again.

The time of potting depends on when the bulbs are available, usually from January to March.

Propagation is by bulblets, seed or 'scales', the scales being pulled off the bulb and inserted upright in peaty compost in warm humid conditions.

NARCISSUS

See Chapter 12, pp 107–14. For cultivation in bowls indoors, see HYACINTHUS

NERINE

Nerine bowdenii, the Jersey Lily, is a hardy plant which makes a good subject for pots if grown four to a 6in pot. Pot the bulbs in March in a well-drained soil with plenty of humus and increase the watering gradually until the leaves die down in late summer, when the showy pink flowers appear.

Hybrid forms are also available for greenhouse cultivation and these should be potted in the same way in August and kept almost dry in light airy conditions until they start into growth. Watering should then be increased until flowering is over and the leaves start to turn yellow, when the plants should be dried off and stored until they are restarted into growth. Repotting is needed only every three or four years and propagation is by offsets then.

POLIANTHES TUBEROSA (Tuberose)

This once popular plant, with extremely fragrant pure white flowers on 3ft stems, is grown from bulbs potted up in early spring. One bulb will go in a 4in pot or three bulbs in a 6in pot, and a suitable compost to use is 4 parts soil and 1 each of peat and sand. Good drainage is important. After potting, place the plants in a temperature of at least 60° F (16° C) and keep them almost dry until growth starts, when plenty of water and light should be given throughout the summer.

The bulbs, which are best discarded after use, are imported and propagation is not carried out in Britain.

SMITHIANA

Showy perennial plants for the warm greenhouse, with large velvety leaves and spikes of tubular flowers, usually in some shade of orange, yellow and red. Pot the tubers from early spring onwards, one to a 5in pot in a compost of 2 parts peat and 1 each of soil and leaf-mould, with a little coarse sand added. Place the plants in a temperature of 55° F (13° C) and keep them just moist, in a shady place. Water more freely and give an occasional feed as soon as the new growths are about 4in high. After the plants have flowered, store them on their sides, without water, through the winter. Propagate by seeds, cuttings and division.

TULIPA (Tulip)

See Chapter 12, pp 107–14. For cultivation in bowls indoors, see HYACINTHUS

VALLOTA SPECIOSA

Scarborough Lily is a showy plant with large funnel-shaped flowers of bright scarlet in August and September. Pot the bulbs singly in well-drained pots just large enough to take them comfortably and keep them almost dry until growth starts. A compost of equal parts soil, peat and sand is suitable.

Water and feed freely during the summer and then keep the plants barely moist through the winter.

Propagation is by offsets when repotting, which is needed only every three or four years.

Chapter 17

Cacti and Other Succulents

The cacti and other succulents are a study in themselves and in a general work such as this it is possible to give only the briefest outline of these plants and their cultivation. However, this should be sufficient to enable the amateur to grow many of the more common kinds successfully.

The term 'succulent' is applied to those plants with fleshy leaves or stems, or both, which enable them to store enough moisture for them to stay alive when they are unable to obtain sufficient water otherwise. Naturally, the degree of succulence depends on the type of plant. Even the common geranium (zonal pelargonium) is, strictly speaking, a succulent as its thick fleshy stems enable it to store enough moisture to stay alive almost throughout the winter. The term is more correctly used for a far more specialised type of plant which is particularly adapted for the storage of water or for the prevention of its loss through transpiration (the process by which leaves give off water as vapour through pores or 'stomata' in the leaf surface). It is usually the peculiar form which this adaptation takes that gives the plants their often curious beauty. In many cases, particularly among the succulents other than cacti, this beauty is still further enhanced by the attractive 'bloom' or waxen covering which protects the plant from excessive transpiration.

The plants vary considerably in size, from the tiny 'mimicry' plants, resembling pebbles, of the genus *Lithops*, to some of the agaves and larger cacti which reach tree-like proportions. Obviously, these larger plants are more suited to outdoor cultivation where climatic conditions allow, but most of them can be grown successfully in pots under glass at least for several years, although in this case there will be little chance of seeing them in flower.

Many of the smaller succulents, on the other hand, flower quite freely, and even among the cacti, which most people may not think of as flowering plants, there are many that are worth growing for their flowers alone. If for no other reason, these flowering kinds make excellent subjects for the greenhouse or indoors, and some of the most suitable will be found in the list at the end of this chapter. They have a further advantage in that, like all succulents, they need far less watering than most greenhouse plants. They make ideal subjects, therefore, for those gardeners who cannot always be at hand to look after them; in fact, they are probably the most labour-saving of all greenhouse plants.

Characteristics of the cacti

The members of the botanical family Cactaceae, all of which are only native to the American continent, provide a large proportion of the succulents in cultivation and they differ from all the others in the possession of special organs known as areoles. It is from these areoles, which are small cushion-like projections generally covered with hair, wool or bristles, that the spines are produced and these spines are themselves an important feature of most cacti, often contributing greatly to the identification of the different species.

Another feature of the cacti is that, with the exception of the genus *Pereskia*, they have no leaves but merely swollen stems, which may be spherical, cylindrical, columnar or flattened in shape. Often these stems are ribbed as in *Echinopsis*, or furnished with tubercles (small nipple-like projections) as in *Mammillaria*, or divided into segments as in *Opuntia* and *Epiphyllum*, but on all the species in any one genus the position of the areoles and spines remains the same, on the ribs, tubercles or flattened stem surface as the case may be.

Desert and epiphytic cacti

The cacti are divided mainly into two types: the so-called desert ones, which are more correctly natives of semi-desert and prairie areas; and the epiphytic ones, which are found in the tropical forests of Central and South America. Like all epiphytes, the plants in this latter group live in trees but are not parasites: their nourishment is gained not from the tree itself but from the pockets of fallen leaves, moss and other humus that collect among the branches. Perhaps the best known example of this group is the Christmas Cactus *Schlumbergera bridgesii*, syn *Zygocactus truncatus*, which is commonly grown as a house plant for the sake of its long red flowers, produced at the ends of the flattened segmented stems in winter.

Cultivation of the desert cacti

The usual way of starting with cacti is by buying young plants in small pots. The plants can be left in these pots until their overall diameter, including the spines, is greater than that of the top of the pot, when repotting will be needed.

The treatment of these young plants is much the same as for the older ones, which means that from April to September they should be given liberal watering (a common mistake, resulting in poor and slow growth, is to keep the plants too dry during this period). From September onwards, however, watering should be reduced gradually until by early winter the plants are being given hardly any water at all, no more, say, than a good soaking every few weeks, until March. Daily overhead syringing will then start them into growth again, when normal summer watering can be resumed. Plenty of light and air are needed at all times and during the winter the plants must be kept safe from frost but not too warm.

When repotting is needed, it should be done when the plants are starting into growth in spring. A suitable compost to use is the J.I.P1 plus one-sixth of its volume of coarse sand and if possible a little crushed brick and charcoal as well. Perfect drainage is essential, but the old idea of filling the pot to about one-third of its depth with crocks is not a good one, as this large amount of space will be far better occupied by compost. All that is needed, therefore, is one good crock placed hollow side downwards so that it covers almost the whole of the base of the pot; above this can go some of the rough sievings from the compost. Potting is then done in the usual way, as described in Chapter 5, but as the spines often make it difficult to work the soil down with the fingers, a small potting stick will be helpful.

Cultivation of the epiphytic cacti

Coming from a quite different habitat to that of the desert cacti, these obviously need rather different treatment, their main requirement being an open, porous and lime-free compost consisting largely of leaf-mould or peat, some soil and a little coarse sand. Shade from full sun is needed in summer, when a fairly humid atmosphere should be maintained; but the main difference between these epiphytes and the desert types lies in the watering. Unlike the latter, the epiphytic cacti must never be allowed to dry out completely; even in winter, when they must be kept safe from frost, they should be watered at least as much as the general run of more ordinary plants. During the summer, or up to the time of flowering in the case of those which flower in winter, they will on the other hand take plenty of water, but when the flowering period is over, it is usual to rest them a little by keeping them slightly drier for a month or so.

Cultivation of other succulents

These vary so much that it is almost impossible to generalise. However, most of them will do well in the J.I.P1 or a mixture of equal parts soil, peat and sand provided watering is attended to properly; in most cases this means giving the plants plenty of water while they are growing and very much less when they are resting. The difficulty lies in knowing when they are at rest and when they are not, for they vary considerably in their season of growth. Some of them, such as the haworthias, gasterias, aloes and South African crassulas, make most of their growth during the winter, when they need fairly liberal watering and a moderate amount of warmth. The 'mimicry' or 'pebble' plants, such as lithops and conophytum, are particularly difficult subjects in respect of their growing and resting periods, but information on this point can be obtained from specialist books on the subject. With most of the easier succulents, however, there is little risk of doing much harm by watering them at the wrong time, and if a plant does begin to look unhappy it is often possible to rectify the trouble simply by giving less or more water according to whether the plant has been kept wet or dry up to then.

Propagation of cacti and succulents

Many of these plants can be raised readily from seed, but unless the seed has been obtained from a reliable source, there is always the risk of producing hybrids rather than true species. Many general seed catalogues offer only mixed hybrids and the seed of definite species needs to be obtained from a specialist grower; or it is possible to save seed from your own plants, a process which is usually described in most books dealing with cacti. The method of sowing the seed is very similar to that described in Chapter 3, but a more suitable compost to use consists of the normal potting mixture surfaced with a layer from which the coarser material has been sieved out.

A more usual way of propagating the different species is by offsets, or by leaf or stem cuttings. In most cases the type of plant will itself indicate the best method to use, for example those cacti which produce offsets or stems that fall off and just ask to be rooted. Some of the easier succulents too will shed leaves or portions of stem that will almost root where they fall, while on some of the cacti with segmented stems, each segment can be used as a cutting.

All these methods of propagation are carried out in much the same way as described for cuttings in Chapter 4, but there are one or two minor differences. Neither leaf nor stem cuttings, for instance, should be placed in close warm conditions, as this may result in their decay. Instead they should be rooted on the open greenhouse bench with just enough warmth to keep them safe from frost, although even this is seldom necessary as propagation is carried out usually in spring or summer. Some cuttings may exude a white juicy substance called latex, and these should be either dried for a few hours before being inserted or dipped in water and then dried. With all cuttings and offsets, watering must be done sparingly, the aim being to keep the rooting medium barely moist.

SOME FREE-FLOWERING CACTI

APOROCACTUS FLAGELLIFORMIS (Rat's Tail Cactus)

An epiphyte with long drooping stems carrying slender red flowers. Useful as a basket plant.

ASTROPHYTUM

A family of mostly globular plants with yellow flowers. Propagation only by seed, which germinates quickly.

CHAMAECEREUS SYLVESTRII

Orange-scarlet flowers produced freely in summer if the plant is kept dry and cool in winter.

ECHINOPSIS EYRIESII

There are some poor strains of this plant which do not flower well but a good specimen should produce magnificent tubular white flowers each several inches long. This plant needs a rather richer and heavier soil than most cacti.

EPIPHYLLUM

The plants commonly grown under this name are mostly hybrids, all of which make excellent and very showy plants for the greenhouse or indoors. The large flowers, in a wide range of colours, are produced usually in early summer and propagation is effected easily by cuttings of the much branched three-angled stems. Treatment as for epiphytic cacti.

GYMNOCALYCIUM

A family of easy-to-grow plants with pink or white flowers. They do best in half-shade in summer and are raised easily from seed.

HAMATOCACTUS SETISPINUS

Yellow flowers, produced freely even on quite young plants. Full sun. Propagation by seed.

LOBIVIA

Most species in this family do well in full sun but the scarlet-flowered *Lobivia cinnabarina* prefers half-shade. Propagation by seed or by offsets.

NOTOCACTUS

All the species in this family are yellow-flowered and all need copious watering and full sun in summer. Propagation by seed or by offsets.

PARODIA NIVOSA

A tiny plant about 2in across, with bright scarlet flowers. Propagation only by seed.

REBUTIA

A family of easily grown plants which flower usually when quite young, often in about two years from seed. *Rebutia miniscula*, with scarlet flowers on a small round plant, is the most popular species.

SCHLUMBERGERA

See ZYGOCACTUS. It much resembles this except in its flowering period which gives it its common name of Easter Cactus.

ZYGOCACTUS TRUNCATUS (now correctly *Schlumbergera bridgesii*)

Perhaps the most widely grown of all cacti, this is a popular greenhouse and indoor plant which produces long red flowers about Christmas. It should be grown in the same way as other epiphytic cacti, but it needs to be kept quite cool in the autumn to initiate the flower-buds. Water it well when it is in bloom.

OTHER GREENHOUSE SUCCULENTS

AGAVE

These handsome American plants, with stiff, fleshy, pointed leaves splaying outwards from a central crown, are far too large for the greenhouse when they are mature, but young plants can be grown inside for several years. *Agave americana* and its variegated forms are the ones most commonly grown, but for the greenhouse a more suitable species is *A. victoria reginae*, a dwarf form with dark green leaves terminating in a sharp black spine and with white stripes along the sides and edges of each leaf.

ADROMISCHUS

A family of dwarf plants, mostly forming small clumps of fleshy leaves. *Adromischus maculatus* is one of the best, with thick wedge-shaped leaves spotted reddish-brown.

ALOE

Like the agaves, most of the plants in this mainly African family grow to a large size, but several do well in pots for several years. There are both tree-like and stemless species, but it is the latter that are grown chiefly, *A. variegata*, the Partridge-breasted Aloe, being particularly popular with its fleshy pointed leaves which are triangular in section and beautifully barred with transverse white bands. Established plants also produce their green-tipped red flowers quite freely.

BRYOPHYLLUM

The species most commonly grown is *Bryophyllum diagremontanum*, which is interesting mainly for its multitude of tiny plantlets that develop on the leaves and fall off eventually and root. It is a very easy plant to grow, 2ft or more high, with serrated pale green leaves spotted purple-red. Grown in a rich moist soil, it should produce heads of reddish flowers in winter.

CONOPHYTUM

Several species of this curious member of the Mesembryanthemum family are available. All of them form pebble-like plants, each 'pebble' consisting of two closely united leaves separated by a slit through which the daisy-like flower emerges. After flowering, a new plant forms inside the old, which shrinks to a thin protective skin. During this period, usually from the end of December onwards, no water should be given; when the skin splits, watering should be increased gradually, although the plants must never be kept wet. Repotting is seldom needed but when necessary it should be done in August, using the same compost as for the desert cacti.

CRASSULA

This family covers many different plants, but most of those grown are the South African species. They are usually easy to grow in any good well-drained soil. In this they should be kept moderately moist in summer and wetter in winter, when they will appreciate comfortable warmth. The most popular species are the shrubby *Crassula arborescens*, with fleshy rounded leaves, and *C. falcata*, a shrubby species which with its bright red flowers makes a good house plant.

EUPHORBIA

This enormous and varied family includes a number of interesting succulent plants, some of them resembling cacti. *Euphorbia obesa*, of ribbed rounded shape and tartan-like colouring, is one of the most popular but there are many others that are well worth investigating.

FAUCARIA

Almost stemless members of the Mesembryanthemum group, with short thick leaves edged with soft 'teeth'. *Faucaria tigrina* (Tiger's Jaws) is the most popular, with the same yellow and almost stemless flowers typical of the other species. The main growth of these plants is achieved in the autumn, when they need the most water.

FENESTRARIA

This genus is also included in the Mesembryanthemum group and covers only two species, *Fenestraria aurantiaca* and *F. rhopalophylla*. These are rather similar 'pebble' or 'window' plants, the latter term referring to the translucent upper surface of the stump-like leaves. Cultivation is as for *Conophytum*, but the plants should be kept dry all winter and watered only sparingly from March to September.

GASTERIA

An interesting family of dwarf, almost stemless plants with fleshy and often beautifully marked leaves opposite each other in pairs which develop one above the other, like the pages of an open book. *Gasteria disticha*, with dark green leaves spotted white, is one of the most common but all the species and hybrids are relatively easy to grow.

GIBBAEUM

This is another genus in the Mesembryanthemum group, with most of the plants forming very short-stemmed clumps of thick fleshy leaves united in pairs. One leaf in each pair is often longer than the other. As the species vary in their periods of growth, they need to be studied individually to find the correct watering period for them and then at all other times they should be kept quite dry.

HAWORTHIA

This large family includes some very varied plants but the general shape is that of a rosette, which on some species is carried on a short stem. As growth takes place chiefly in winter, when the plants need to be merely safe from frost, watering is done mainly then, with less water being given in summer. *Haworthia margaritifera*, with white pimples on its dark green leaves, is one of the most commonly grown.

KALANCHOE

Several species in this family, such as *Kalanchoe flammea* and *K. tomentosa*, are grown frequently in collections of succulents, but the most common is *K. blossfeldiana*, a popular house plant with showy heads of red flowers in early spring. This and the other types are all fairly easy to grow from seed or cuttings.

KLEINIA

Unlike the other succulents in this list, the *Kleinias* (sometimes included in the genus *Senecio*) are members of the botanical family Compositae, characterised by daisy-like blooms. Their flowers are, however, seldom produced in cultivation although the best known species, *K. articulata*, sometimes produces its small white flowers above the curious jointed stems, which vary considerably in shape according to the amount of light they get.

LAMPRANTHUS

Originally forming part of the Mesembryanthemum family, the plants in this genus are semi-shrubby in character, one of the best being *Lampranthus spectabilis*, a prostrate plant with large purplish-pink 'daisies'. All the cultivated species are better planted out for the summer than grown in the greenhouse, and propagation is effected easily by cuttings in late summer.

LITHOPS

This family of 'living stones' are among the most fascinating of the succulents. The body of each one consists of a pair of closely united leaves, although these would hardly be recognised as such with their pebble-like appearance, flat tops, and in some cases characteristic 'bird's egg' markings. The flowers, white or yellow, are produced from July to November. Cultivation is largely as for CONOPHYTUM (see pp 176–7), but the resting period is from December to May, when no water should be given until the leaves have shrunk to a dry skin.

MESEMBRYANTHEMUM

This name has now gone out of use for practically all the plants it covered originally, and even the popular annual bedding plant *Mesembryanthemum criniflorum* (the Livingstone Daisy) is now correctly known as *Dorotheanthus criniflorum* or *D. bellidiflorus*—the two names appear to be synonymous. For genera previously known as *Mesembryanthemum* see CONOPHYTUM, FAUCARIA, GIBBAEUM and LITHOPS.

ROCHEA

Of this small genus of four species the only one commonly

grown is *Rochea coccinea*, a showy flowering plant sometimes sold as *Crassula coccinea*. The heads of red tubular flowers are produced in summer above erect stems covered with short thick leaves which are set close together in pairs arranged in a criss-cross fashion. Propagation is effected easily by cuttings of the unflowered stems taken in spring or summer.

SANSEVERIA

See list of foliage plants in Chapter 15, pp 139–54

STAPELIA

A genus of which the plants are probably grown mainly as a curiosity, for they are hardly attractive—as their common name of Carrion Flower implies. The most common species, *Stapelia variegata*, has thick toothed stems from the base of which the yellow and brown foul-smelling flowers, in shape resembling five-pointed stars, are produced.

Chapter 18

Greenhouse Plants Indoors

Whenever there is a good show of plants in the greenhouse there is no reason why a few of the best should not be taken indoors. This is not without its risks, however, for not only is the atmosphere indoors very different from that of the greenhouse but there are likely to be draughts, dust and possibly even gas fumes sometimes. Many of the tougher greenhouse plants will stand up to these difficult conditions, but the more tender ones, particularly those needing a warm moist atmosphere, may soon suffer if they are left indoors too long. A good method is to bring a plant indoors for a day or two, then to return it to the greenhouse and exchange it for another, taking care each time to make conditions indoors as similar as possible to those of the greenhouse.

Apart from this, the treatment of greenhouse plants indoors is much the same as that of the general run of house plants. Those needing a humid atmosphere, for instance, should be stood on a moist base such as a tray of wet gravel or peat to create the necessary 'micro-climate' around them, and they should never be placed in full sun. Most plants, in fact, apart from a few such as geraniums and cacti, do not like to be baked all day in full sun or to be given a dry atmosphere with inadequate ventilation as well. During the summer many plants will do better in a north-facing window than in a more sunny one, but if they have to be kept in the latter they should at least be shaded by a curtain, say, during the hottest part of the day. Shade-loving subjects on the other hand are easier to accommodate but even these should, as far as possible, be kept out of draughts and safe from very fluctuating temperatures.

Sponging the foliage of large-leaved evergreen plants is another point that should not be neglected, as otherwise they will soon collect enough dust to prevent their leaves from functioning properly. However, if they should get into this unhappy state, an hour or two out-of-doors in mild gentle rain will soon clean and freshen them up, but make sure they are safeguarded against being blown over by the wind.

Raising young plants indoors

With the aid of a modern electric propagator, most seeds and all types of cuttings may be coped with successfully indoors in much the same way as in the greenhouse, and even without one, good results can be obtained with most plants. Seeds for instance will usually germinate reliably over a radiator or even in an airing cupboard provided they are kept moist, but in either case they must be given more light immediately they are through. The snag with seedlings raised indoors, lies in what to do with them when they are ready for pricking out, for unless only one or two sorts are grown in very small quantities, they will take up a good deal of room then. For this reason the times of sowing should be adjusted so that when the seedlings are pricked out they may be placed safely in a cold frame without any risk of damage by frost. Even then it is advisable to cover the frame with mats or sacking on cold nights, at least until the plants have got used to the change of temperature.

Cuttings are rather easier to deal with, as they are normally taken only in quite small quantities, often no more than two or three of any one kind being taken. The simple method of rooting shoot cuttings in water has been described already in Chapter 4; another and almost equally simple method is to insert them round the edge of a pot of well-drained sandy soil and merely stand them in a north-facing window, where they should be left to become almost dry between thorough soakings. This should be done preferably in summer, as most softwood cuttings root most readily then, but even in winter good results can often be obtained provided there is adequate warmth and the cuttings are given as much light as possible. Another method still is to insert the cuttings in a pot of soil and to enclose the whole thing in a plastic bag. They will need no water in this until they are eventually rooted and removed from it. This method works particularly well for evergreen shoot cuttings and for those cuttings requiring a humid atmosphere, such as leaf cuttings of *Begonia rex*, but it should not be used for 'dry' subjects such as geraniums which need plenty of fresh air.

Some of the climbing or trailing plants, such as the ivies, may also be layered easily indoors. The best way to do this is to stand a small pot of soil beside the plant and to peg down a shoot in it so that one of the leaf-joints is firmly in contact with the soil. A new shoot will then emerge from the leaf-joint and eventually the new plant can be severed from the parent one and grown on just as it is.

Chapter 19

Some Popular Greenhouse Plants

In the previous chapters I have dealt with a number of plants that fall into definite groups according to their main characteristics or habit of growth, but there are many others that are commonly grown. The following list covers most of those generally grown by the amateur and also a few that are not quite so well known. Given normal greenhouse treatment none of them should present much of a problem, so my object has been to stress the more important points of cultivation and to indicate where necessary the possible causes of failure.

ABUTILON (Indian Mallow)

Half-hardy shrubs with bell-shaped or lantern-like flowers in summer and autumn. Several fine hybrids are available and these, together with such species as *Abutilon megapotamicum* (crimson and yellow) and *A. striatum thompsonii*, should be kept safe from frost and almost dry in winter. Cut back straggling shoots in February and propagate by cuttings of the old wood in heat in autumn or spring.

AECHMEA

See BROMELIADS

ANTHURIUM (Flamingo Flower)

Anthurium scherzerianum is a popular house plant grown for the large and vividly coloured spathe which surrounds the insignificant flowers. It needs a humid atmosphere and ample warmth, with liberal watering and light shade in summer and less of both in winter. Propagation by division is best carried out in warm conditions in January, when repotting should also be done. A rough open compost consisting largely of humus with plenty of coarse sand, crocks and charcoal mixed in for drainage suits this plant, which should be potted so that the base of it stands well above the pot rim on a mound of the compost.

APHELANDRA (Zebra Plant)

This is another popular house plant, grown usually in the form of *Aphelandra squarrosa louisae*, with large ivory-veined leaves and a prominent four-angled spike of bracts, which should be removed as soon as it has faded. This plant should be kept just moist in winter in a temperature of about 50° F (10° C) and then repotted in spring, when the new shoots may be used as cuttings. Do not feed until the flower-buds appear in late summer, otherwise the plant may produce leaves only.

AZALEA

The showy *Azalea indica* (*Rhododendron simsii*) is one of the most popular Christmas-flowering pot plants, needing only to be kept just moist in cool conditions. Lime-free water (rain-water if necessary) should always be used for it. When the faded flowers have been removed in spring, the plant should be grown on indoors until June; it should then be placed outside and kept watered and fed until it is brought in again in the autumn. Cuttings of the young shoots will root in close conditions in summer. When they have rooted, they should be potted up in a lime-free mixture of peat and sand.

BELEROPONE GUTTATA (Shrimp Plant)

Evergreen shrub with an almost continuous display of the decorative brownish-pink bracts that give it its common name. This is a quite popular house plant needing a winter temperature of 50° F (10° C). To produce cuttings some of the main stems should be cut back in spring, when repotting should also be done. Too much heat will result in bracts of a poor colour.

BILLBERGIA
See BROMELIADS

BOUVARDIA (Jasmine Plant)

Once very popular, these Mexican evergreen shrubs, with showy and sometimes very fragrant flowers, are seldom seen these days although it is possible to obtain a limited selection from a few nurseries. The plants need liberal feeding and watering in summer, when they should also be pinched out frequently to induce a bushy habit, and this should bring them into bloom by autumn or winter. After flowering they should be rested by being kept dry; then in spring they should be cut back before they start into growth again. Repotting is best done in spring, when a well-drained compost of soil, leaf-mould and well-rotted manure should be used. Propagation is effected by cuttings of the young shoots, taken about 2in long at repotting time. These will root readily in a warm close frame provided their leaves are not wetted.

BROMELIADS

This is a general term for the members of the botanical family Bromeliaceae, some of which are now commonly grown as indoor or greenhouse plants. With the exception of a few genera such as *Billbergia* and *Cryptanthus*, which are terrestrial plants needing more or less ordinary treatment, most of those in cultivation are epiphytes growing naturally on trees in the tropical forests of America, and their cultivation is thus different from that of the general run of plants.

The main characteristic of these epiphytic types, which include *Caraguata*, *Guzmannia*, *Neoregelia*, *Nidularium*, *Tillandsia* and *Vriesia*, is the central 'urn' or 'vase' formed by the rosette of leaves. As long as this is kept filled, very little watering of the actual soil is needed, particularly in winter.

After the plants have flowered, the original rosette of foliage dies and its place is taken by offsets, which may be removed and potted up, preferably in August. A well-drained compost consisting of leaf-mould, sphagnum moss and peat should be used and each offset should be placed in as small a pot as possible and grown on with plenty of warmth, light shade and careful watering. Once the offsets are established, they can be given more light and less warmth. It is important to guard against over-potting: repotting should not be done until it is absolutely necessary, usually when the plant has become top-heavy for its pot.

BRUGMANSIA

See DATURA

CALCEOLARIA

These showy plants, with large pouch-shaped flowers in a wide range of colours, are ideal for the amateur grower. Seed sown about July will produce flowering plants for the spring. During the winter it is important to keep them safe from frost, with thorough waterings each time the compost almost dries out.

CAPSICUM

Small ornamental plants grown for their colourful fruits. For treatment see SOLANUM CAPSICASTRUM.

CARAGUATA

See BROMELIADS

CELOSIA (Cockscomb)

Half-hardy annual. Sow in brisk heat in March and grow on in 5in pots of rich soil, with plenty of warmth and water, until the plants are a good size, when they should be kept drier to produce the large and brilliantly coloured heads of bloom.

CINERARIA

One of the most popular of spring-flowering plants, this can be grown easily with the minimum of heat. Sow the seeds in June or July and grow on into 5in pots, with shade from full sun. Over-watering in winter is a great danger with cinerarias: to be on the safe side it is best to let the plants flag a little from dryness before soaking them thoroughly. Beware of greenfly!

CRYPTANTHUS

See BROMELIADS

CUPHEA IGNEA (Mexican Cigar Plant)

An attractive evergreen subshrub, 12in or so high, with narrow scarlet flowers tipped with black and white. It is suitable for both the greenhouse and bedding and is raised easily from seed sown in heat in spring or from cuttings taken at the same time. Young plants should be pinched back once or twice to induce a bushy habit.

DATURA (Angel's Trumpet)

The shrubby species make striking and unusual plants for tubs or large pots in the cool greenhouse, but young plants do not flower as freely as old ones. *Datura suaveolens*, with white trumpets up to 12in long on a plant eventually several feet high, and the rather similar *D. cornigera* (known as *Brugmansia knightii* in its double form) are the ones usually grown and both are raised from cuttings in heat in spring or autumn. A compost of soil, peat, sand and well-rotted manure in equal parts suits them and during the summer the plants are best stood outside. Plenty of water is needed in summer, much less in winter, when the plants should be kept cool but safe from frost. Pruning is done after flowering, by cutting hard back the shoots of the current year.

EUPHORBIA PULCHERRIMA

See POINSETTIA

EXACUM AFFINE

This 6in high biennial makes a quite pretty pot plant, with mauve-blue flowers all summer. Sow the seed in late August and grow the plants on through the winter in at least 50° F (10° C), with only enough water to keep them barely moist. Perfect drainage is essential to prevent damping off. In spring move the plants on into 5in pots and give them more water in a humid atmosphere with shade from full sun.

FUCHSIA

This invaluable plant has now regained much of the popularity it enjoyed in Victorian times and many fine cultivars, both single and double-flowered, are offered by specialist growers. To keep these true to name (ie to reproduce named varieties identical to the parent plant) they are raised invariably from cuttings, which are taken about 2in long early in the year, after the old plants have been cut back and restarted into growth. These cuttings will make good flowering plants during the summer, then in the autumn they should be stored in a frostproof place, with only enough water to prevent them from drying out completely.

In summer liberal watering, light shade and a fairly humid atmosphere provide the best growing conditions, but the plants are not very fussy and the only real trouble that is likely to be encountered is the shedding of the flowers from indoor plants, caused usually by lack of water and a too dry atmosphere.

The trailing forms are particularly useful for hanging baskets and other plant containers, while both the erect and trailing types can be used to form 'standards', each consisting of a bushy head at the top of a bare stem which may be as much as 3ft high. To produce a standard, a strong erect cutting is potted up and tied to a cane to be kept straight; it is then grown upwards, and all sideshoots from it are removed until the required height is reached. Any leaves actually growing on the stem of the cutting, however, should be retained. When the stem has reached the necessary height, the sideshoots are allowed to grow above this point and if these are stopped once or twice there will soon be a good head of bloom.

GARDENIA

The fragrant white flowers of this evergreen shrub make it an ideal greenhouse plant, which with adequate heat can be brought into bloom at any time of the year. *Gardenia jasminoides* is the species most commonly grown and cuttings of this should be taken in brisk heat in a propagating frame early in the year. They should then be potted up into a rich humus soil over good drainage and grown on with plenty of warmth, water and humidity. To produce bushy plants, the cuttings should be stopped at a few inches high, and when they have flowered, eventually new plants should again be raised from the cuttings as young plants raised annually flower more freely than old ones.

GUZMANNIA

See BROMELIADS

HIBISCUS

The tender *Hibiscus rosa-sinensis* has become increasingly popular in recent years and its large colourful blooms make it an invaluable addition to the greenhouse and indoor plants. It is best grown in a rough peaty soil, with plenty of water in summer, less in winter, when it should be kept in a temperature of about 50° F (10° C). To restart it into growth, cut it back lightly in spring and give more water and warmth; then throughout the summer it should be grown on in full sun. Propagation is by cuttings of the ripe wood in autumn, in a close frame with bottom heat.

HYDRANGEA

The popular hydrangeas brought at the florist's consist of the hortensia varieties of *Hydrangea macrophylla*. They have only sterile flowers and must therefore be propagated by cuttings. These may be taken from March to August, the earlier ones being stopped once to produce plants with several heads of bloom the following summer and the later ones being grown on without stopping to form single-headed plants.

On hydrangeas the buds that will develop into flowers the following year are formed in the autumn; to ensure this the plants should be stood outside and kept moist until the leaves fall, when they should be brought inside and kept cool and almost dry throughout the winter.

To retain the colour of the blue varieties, the plants must be grown in lime-free soil and fed with a proprietary colouring agent.

IMPATIENS (Busy Lizzie)

See list in Chapter 13, pp 115–25. Plants grown indoors need copious supplies of water in summer

MIMOSA PUDICA

The Sensitive Plant is so-called because its leaflets droop and come together at the slightest touch. It needs a high temperature and plenty of humidity in summer and is best discarded at the end of the first season, with new plants being raised from seeds in spring or from cuttings in early autumn. A temperature of at least 60° F (16° C) is needed for both seeds and cuttings.

NEOREGELIA

See BROMELIADS

NERIUM

Commonly known as the oleander, *Nerium oleander* makes a fine evergreen flowering shrub for the cool greenhouse, with flowers in some shade of white, pink or purple in summer. It needs a compost of equal parts soil and well-rotted manure. Make sure that the plants get plenty of water, light and air in the summer, so that the shoots are thoroughly ripened to the flowering stage. Propagation is by cuttings, which root easily in water. This plant is very poisonous in all its parts.

NIDULARIUM

See BROMELIADS

PELARGONIUM

The zonal pelargonium (geranium) has been dealt with already in Chapter 13, but there is another sort which is commonly grown as a greenhouse or indoor plant. This is the Regal pelargonium, with larger and more showy flowers, and its treatment is rather different from that of the zonals.

The plants are usually purchased in bloom in early summer. Once flowering is over they should be placed outside in full sun and kept moist until August, when cuttings should be taken. These consist of ripe shoots removed at about 3in long. After trimming them off beneath a joint, insert them round the edge of a 5in pot of well-drained sandy soil and place them in the greenhouse or frame. When rooted they should be potted up into 3in pots and eventually into 5in ones for flowering during the summer after the cuttings were taken. For early flowering, the plants should be stopped about the end of November and for later ones this should be followed by a second stopping in February. Rather more warmth and water are needed in winter for this type of pelargonium and the plants should never be allowed to dry out then.

The original plants may be kept for a second year. After the cuttings have been taken, the plants should be laid on their sides in full sun and kept dry until the shoots are quite firm and brown. Frequent syringing should be carried out to induce the plants to start into growth; they should then be shaken out of the old soil and repotted into the same size of pot and grown on in the same way as the cuttings throughout the winter.

POINSETTIA

With its large scarlet bracts resembling flowers, this plant—correctly, *Euphorbia pulcherrima*—has become quite popular as an indoor plant during autumn and winter. The treatment at that time consists of giving the plant a temperature of at least 60° F (16° C) and watering it thoroughly each time the soil is quite dry on the surface. When flowering is over, it should be kept quite dry and safe from frost until it starts into growth again about May, when it should be cut back and repotted. The cut-off portions may be used as cuttings. Plenty of water and warmth are needed in summer, but in autumn it is important not to expose the plant to artificial light, as this may delay or prevent flowering.

PRIMULA

This family includes two invaluable plants for the greenhouse and indoors: one is the almost perpetual-flowering *Primula obconica*; the other, the spring-flowering Fairy Primula *P. malacoides*. Both plants are grown from seed, the former being sown usually in heat in spring and the other without heat about June, preferably in a north-facing cold frame. Pricking out and potting on *P. malacoides* then follows the same pattern as for other plants, but with *P. obconica* better results are obtained if the plants are pricked out 3in apart into 3in deep boxes, so that the normal 3in pot stage is eliminated. Liberal watering and light shade are needed in summer. In winter *P. malacoides* should be kept just safe from frost, with plenty of ventilation, and *P. obconica* should be kept preferably in a temperature of at least 55° F (13° C) in which it will take copious watering. Both plants are discarded usually when flowering is over.

SAINTPAULIA IONANTHA (African Violet)

In recent years this plant has become a great favourite. It is grown easily in summer, when it needs plenty of warmth, water and humidity, with shade from full sun. It may be raised from seed sown in good heat in spring, or established plants may be propagated by leaf cuttings as described in Chapter 4. It is not an easy plant to keep through the winter, although it will often do well at that time on a window-sill in a warm kitchen.

SCHIZANTHUS (Butterfly Flower or Poor Man's Orchid)

This showy annual, with attractively marked flowers in a wide range of colours, is always popular with amateurs, no doubt because it can be raised very easily from seed. For small plants, seed may be sown in heat in spring, but better results and larger plants are obtained by sowing in August or September and keeping the resultant plants cool, airy and almost dry in full light throughout the winter. During this winter period the plants should be in 3in pots; then in spring they should be potted on into 5in or 6in ones, a stopping being given when they are a few inches high. Staking is usually necessary.

SOLANUM

The most popular species in this family is *Solanum capsicastrum*, the Christmas Cherry, with large orange and red berries. This is raised easily from seed sown in heat in spring, the seedlings being potted up eventually into 3½in or 5in pots of the J.I.P2. To get a good set of berries, the plants should be plunged to the pot rim outside and sprayed frequently overhead when they are in bloom from June onwards. In the autumn they should be placed in a frostproof greenhouse and kept just moist. This plant is particularly prone to magnesium deficiency, indicated by the yellowing and falling of the leaves; to guard against this, a pinch of Epsom salts (magnesium sulphate) should be worked into the compost for each pot. Propagation may be carried out by cuttings, taken after the old plants have been cut back in spring.

SPARMANNIA AFRICANA (African Hemp)

This plant, with large hairy leaves and heads of white flowers in summer, is often grown as an indoor as well as a greenhouse plant. It needs a peaty soil and should be kept well watered in summer and just moist in winter, when it should be cut hard back about February. It is easily propagated by cuttings in warm close conditions.

STRELITZIA REGINAE (Bird of Paradise Flower)

The large purple and orange flowers, resembling a stylised bird's head, makes this one of the most striking of greenhouse plants. It is best planted out on a well-drained mound of rich rubbly soil in the greenhouse, but it may be grown also in a rather heavier soil in a well-drained 9in pot. Copious watering is needed in summer but very little in winter, when a temperature of 50° F (10° C) is quite high enough. Propagation is by suckers, or by seed, which sets only after hand fertilisation.

STREPTOCARPUS (Cape Primrose)

Sown in good heat in early spring, this showy plant with large bell-shaped flowers will be in bloom by late summer. It needs warm and fairly humid conditions, with shade from full sun.

In winter it should be kept safe from frost and almost dry.

Although the best plants are those raised annually from seed, the streptocarpus may be propagated also by leaf cuttings in the same way as the gloxinia.

TILLANDSIA

See BROMELIADS

VRIESIA

See BROMELIADS

ZANTEDESCHIA AETHIOPICA (Arum Lily)

With its large white trumpet-shaped flowers (spathes), this is one of the most popular flowers of all, both for cutting and for pots. The fleshy rhizomes should be potted up singly into 6in pots in July, using a rich rough soil with perfect drainage, and the plants should then be stood outside and given gradually more water until they are housed in September. In a cool-house temperature of 45° F (7° C), they will flower in spring, but more heat will bring them into bloom earlier. After flowering, they should be placed outside and allowed to die down until they are repotted again in July; alternatively, they may be planted out in the garden from July onwards where if given a rich soil and liberal watering they will make fine plants for potting up in September. Propagation is by suckers when repotting is being done.

Chapter 20

A Greenhouse Calendar

JANUARY

Sowings to make: gloxinias, streptocarpus, cyclamen, tomatoes and French beans; also, in mild areas, lobelia, penstemons, verbenas, *Begonia semperflorens* and *Salvia splendens* for bedding out later. All these need a temperature of at least 60° F (16° C) but antirrhinums, sweet peas, lettuce and mustard and cress may be sown now in cooler conditions.

Take cuttings of Perpetual carnations, Large Exhibition and late-flowering (Christmas) chrysanthemums.

Divide and repot anthuriums and pot new lilies as they become available.

In the heated greenhouse start vines, peaches and nectarines.

Get in hand a supply of peat, sand, potting compost and fertilisers etc ready for use.

FEBRUARY

Sowings: as for January, but those mentioned for mild areas may be sown elsewhere now together with all other summer bedding plants except zinnias, nicotiana and bedding dahlias. *Primula obconica* may also be sown now for greenhouse or indoor decoration with temperature of 60° F (16° C).

If adequate heat is available, start dahlia, gloxinia and begonia tubers, also hippeastrums, vallotas and achimenes.

Cut back stored fuchsias, zonal pelargoniums, coleus and heliotropes and start them into growth.

Take cuttings of decorative chrysanthemums, also Perpetual carnations.

Prune bougainvillea and *Passiflora caerulea* (passion-flower).

Pollinate and disbud early peaches and nectarines.

MARCH

Sowings: as for February, plus bedding dahlias and nicotiana; also schizanthus for summer flowering in pots and *Primula obconica* and *Solanum capsicastrum* for Christmas. Melons and cucumbers should also be sown now.

Take cuttings of fuchsias, zonal pelargoniums, coleus, heliotropes and dahlias, and start dahlias, gloxinias, begonias etc into growth if this has not already been done.

Continue to take cuttings of Decorative chrysanthemums and stop young Perpetual carnation plants at six pairs of leaves.

Start cacti and ferns into growth and repot if necessary. Most indoor plants may also be repotted now.

Start vines in unheated houses and prune and train those started earlier.

Plant tomatoes in heated greenhouses and thin the fruits on early peaches and nectarines.

APRIL

All the sowings mentioned so far may still be made, but the sooner the better. Zinnias should be sown at the end of the month.

Harden off the earliest bedding plants.

Most cuttings, including hydrangeas, may also still be taken and the various bulbous plants started into growth.

Dry off freesias, lachenalias and cyclamen as they finish flowering but keep other plants just coming into growth well watered now.

Cut back *Hibiscus rosa-sinensis* and start it into growth.

Plant tomatoes, cucumbers and melons in heated greenhouses.

Continue to prune vines and thin the fruits on peaches and nectarines.

MAY

Harden off and plant out all bedding plants except those mentioned below for June planting. Make up hanging baskets.

Chrysanthemums: November-flowering singles, also December-flowering decoratives being grown for second crowns, should be given their first stopping early in the month. The final potting or planting of all chrysanthemums should be completed by about the end of the month.

Dry off and rest arum lilies as they finish flowering.

Plant tomatoes, cucumbers and melons in cold houses.

Start to thin grapes as soon as the berries are about ⅛in across.

JUNE

Harden off and plant out zonal pelargoniums, fuchsias, *Salvia splendens*, *Begonia semperflorens*, heliotropes and dahlias.

Complete the stopping of Perpetual carnations in cold houses by the middle of the month.

Sow cinerarias and *Primula malacoides* for flowering early next spring.

Chrysanthemums: October/November-flowering decoratives should be stopped at the end of the month if they have not already broken. December-flowering cultivars should be stopped at the same time for either first or second crowns.

JULY

Sow greenhouse calceolarias and make further sowings of cinerarias and *Primula malacoides*.

Cut back and repot hydrangeas as they finish flowering, and dry off and rest Regal pelargoniums when these two have finished.

Complete the stopping of Perpetual carnations in heated greenhouses by the end of the month.

Chrysanthemums: early in the month stop the November-flowering singles for the second time.

Keep hippeastrums well watered and fed in full light after they have flowered.

Take leaf cuttings of *Saintpaulia ionantha* (African Violet), gloxinia, streptocarpus and *Begonia rex*.

As the fruit is cleared from peaches and nectarines, give plenty of ventilation and keep the trees well syringed.

AUGUST

Sow cyclamens, schizanthus and mignonette for next year. Repot old cyclamen corms.

Take cuttings of zonal pelargoniums, gazanias and fuchsias (the last-named for growing as 'standards').

Pot up corms of freesias and lachenalias and keep them cool and almost dry in a cold frame.

Repot and start into growth arum lilies and Regal pelargoniums. Cuttings of the latter should be taken now.

Chrysanthemums: as the flower-buds appear from now onwards they should be retained or 'secured'.

SEPTEMBER

Start to pot lilies as they become available.

Pot freesias and lachenalias for succession.

Chrysanthemums: bring pot-grown and 'lifted' plants inside at the end of the month. Continue to secure the buds as they appear and carry out disbudding as it becomes necessary.

Start the planting of hyacinths, narcissi (including daffodils), early tulips and other hardy bulbs in pots and bowls.

Ripen off melons in the greenhouse or frame.

OCTOBER

Pot plants which have spent the summer in a frame must be brought inside now, but leave hydrangeas out until the leaves have fallen.

Bring in perennial bedding plants such as zonal pelargoniums, fuchsias, heliotropes and *Begonia semperflorens* before they are damaged by frost. Dahlias and tuberous begonias can be left until the foliage has been blackened by frost.

Gloxinias and tuberous begonias grown in pots should be dried off gradually ready for storing, the former in a temperature of at least 50° F (10° C).

No further damping down or syringing should be carried out in the greenhouse; instead, use a little heat and give plenty of ventilation to provide a dry buoyant atmosphere.

Bring in freesias and lachenalias as soon as growth is well started; also arum lilies for early bloom.

Continue to plant bulbs in pots and bowls.

NOVEMBER

Prune the greenhouse climber *Plumbago capensis* by cutting the flowered shoots back to about 9in.

Dry off pot-grown fuchsias and store in a frostproof place.

As the first greenhouse chrysanthemums finish, cut them down to a few inches high.

The earliest bulbs in pots and bowls may be brought into mild heat now provided growth is adequately forward.

Close all ventilators during foggy periods and give more warmth and ventilation afterwards.

Bring in pot-grown hydrangeas when their leaves have fallen.

Prune grape-vines as soon as the leaves fall and give plenty of ventilation.

DECEMBER

Take cuttings of Perpetual carnations and give the mature plants maximum light and air.

Continue to bring in potted bulbs as in November.

Keep the best chrysanthemum plants for cuttings and discard the rest.

Clean pots, seed trays etc ready for the coming season, and put soil for seed and potting composts under cover.

Cyclamens in full bloom should be kept fairly warm in full light and given liberal watering.

Prepare seed list for next year.

Index

Specific names are in italics, page references to illustrations are in bold. Where the name of a genus is followed by spp, more than one species is dealt with.

Aralia elegantissima, see Dizygotheca